Soviet Rocketry

Soviet Rocketry:

THE FIRST DECADE
OF ACHIEVEMENT

BY MICHAEL STOIKO

DAVID & CHARLES: Newton Abbot

ISBN   0 7153 5308 X

First published in 1971 in Great Britain
by David & Charles (Publishers) Limited

Copyright © 1970 by Michael Stoiko

Designer   Ernst Reichl

Printed in Great Britain
by Redwood Press Limited Trowbridge Wiltshire
for David & Charles (Publishers) Limited
South Devon House Newton Abbot Devon

Grateful acknowledgement is made to the following sources for use of
material in this book: Academy of Sciences of the U.S.S.R. (Moscow) for
figs. 6, 14, 15, 19–22, 24 from A.A. Blagonravov, *Soviet Rocketry: Some
Contributions to Its History*, 1964; fig. 8 from I. A. Slukhai, *Russian Rocketry, A Historical Survey*, 1965; figs. 4, 9, 10, 13, 16, 18 from V. N. Sokolsky,
A Short Outline of the Development of Rocket Research in the U.S.S.R.,
1960; figs. 1–3, 5 from V. N. Sokolsky, *Russian Solid–Fuel Rockets*, 1963.
    Pravda for figs. 28–30.
    Novosti for figs. 7, 23, 25, 31–69.

To my wife, Margaret Jane Hoehn Stoiko,
for our nineteenth wedding anniversary

Contents

Preface

IN SPITE OF the many Soviet pronouncements that preceded its flight on October 4, 1957, Sputnik 1's launch occurred with the suddenness and surprise of a Pearl Harbor and with the impact of a Hiroshima atomic explosion. Perhaps at no other time in human history had a single event so changed the course of mankind. The streaks of white light and the "beep-beep-beeps" of the first Soviet sputniks in the fall of 1957 served as a warning to the free world of the achievement of Russian science and technology. On October 16, 1957, Richard Nixon, then Vice President, gave one of the sanest Administration reactions to Sputnik 1 when he said, "We could make no greater mistake than to brush off this event as a scientific stunt. . . . We have a grim and timely reminder . . . that the Soviet Union has developed a scientific and industrial capacity of great magnitude."

Starting with the first sputnik, through at least the first five years of the space age, the Soviet Union continued to reap enormous world prestige from its space accomplishments. Worldwide public opinion polls conducted between 1957 and 1962 by the U.S. Information Agency confirmed that, with the successful launching of Sputnik 1 through the launchings of their first Cosmos in 1962, the Soviet Union's prestige abroad increased, while producing a corresponding decrease in United States' prestige. Moreover, each subsequent launch tended to produce shifts in world estimates of the United States' and the Soviet Union's scientific capability, but with the Soviet Union always maintaining a substantial prestige lead over the United States.

The affects of these early Soviet successes were revealed in a 1960 world Gallup Poll that asked the question, "Looking ahead ten years, which country do you think will have the leading position in the field of science?" Of the American respondents, 70 percent named the United States, 16 percent cited the Soviet Union, 2 percent mentioned other countries, and 12 percent did not know. In Europe, only two of the nine countries polled, Greece and West

Germany, assigned a superior position to the United States, and then by a margin of less than 10 percent.

Sputnik's success was not easily achieved. Prior to its launch very little was known about near or far space. The only available information came from terrestrial observations and sounding rockets. In all cases, the data was either incomplete or not accurate enough to allow development of a realistic atmospheric profile.

In the decade that followed, hundreds of Soviet spacecraft, unmanned and manned, have performed many complex missions proving beyond doubt the high state of its technological development. Yet, within that same ten years' time, the American public has learned little about the phenomenal growth of Soviet space power. In fact, the image of presputnik Russia was that it was populated with heavy-bearded Moujikas and with scientists possessing no unique scientific or technological know-how.

Even after ten years, the assessment of Soviet capabilities and intentions continues to be misinterpreted by some of the most knowledgeable individuals, creating unnecessary confusion and apprehension. This situation need not exist, for in the last few years, great care has been taken in both the United States and in the Soviet Union to compile and document factual material on the development of Soviet rocketry and space flight. In the preparation of *Soviet Rocketry* many of these official and authoritative sources were minutely researched.

In selectively compiling notes it was my intention to write a factual book depicting chronologically for the first time the events following the introduction of rocketry to Russia up to the end of the first decade of the space age. The descriptive texts taken from official Soviet sources have been edited only when absolutely necessary in order to retain their authenticity. In specific cases where it appears the Soviet texts have, for one reason or another, "overshadowed" known facts, I have rewritten that portion in an objective manner retaining the historical content of notes and research. And in a comparable manner, source material which originated outside of the Soviet Union was also carefully researched, updated, and objectively presented.

Because a knowledge of the past and present can help us prepare for tomorrow, it is my hope that this account will provide the reader with the information he needs to judge the real importance of the "spectaculars" and the "firsts," as well as the long-term implications of space flight as far as he, his country, and the world are concerned. Then, and only then, can he realistically assess his country's future in terms of his support and aspirations for his nation in space.

A book is almost always the product of the efforts of many people. This one was no exception. Accordingly, I want to thank the many people at the Library of Congress who were so helpful in researching material for this book. In particular, the help of Marvin W. McFarland, Chief of Science and Technology Division; the assistance and cooperation of Analyst Leonard N. Beck and Alexsander Dolgich, Supervisor, of the Aerospace Vehicles and Space Exploration Unit; of Dr. Charles S. Sheldon II, Acting Chief of the Science Policy Research Division; Dr. Ivan M. Soukhanov, Supervisor of the Aerospace Science Unit; and Dr. Joseph G. Whelan, Analyst, Foreign Affairs Division, are especially appreciated and gratefully acknowledged.

Special thanks to Dr. Eugene Emme, NASA Historian, for pulling together many reference sources, and to Janis L. Koch for her accurate translations.

Much of the Soviet descriptive material and many of the photographs used in this book were provided by Novosti Press Agency through the cooperation of Mrs. Gorchakova, Information Department, and Mr. Preferansky, Assistant Commercial Counselor of the Embassy of the Union of Soviet Socialist Republics.

M.S.

Towson, Maryland
January 1970

Foreword

The Space Age began on 4 October 1957, with the launching of Russia's Sputnik 1 — the first artificial satellite. Since then there have been spectacular developments, culminating with the lunar landing by Apollo II in 1969. All the important launchings have been carried out either by the USA or by the USSR, and as yet the two programmes have been undertaken quite separately.

The methods have been somewhat different. The Americans have concentrated upon both automatic probes and manned ventures; the Russians have made no attempt to send men to the Moon, but have sent up the highly sophisticated manned Soyuz vehicles during 1968-70 that made completely new kinds of manoeuvres, rendezvous and crew exchanges; in 1970 Soyuz 9 orbited for a record eighteen days. In 1970 the Soviet Union also landed Venus 7 on the surface of the planet, while Lunokhod 1, capable of guided movement on the lunar surface, sent back valuable data from areas too dangerous for manned landings. There is a difference, too, in another way. All the American lunar vehicles have been despatched in a blaze of publicity, and the whole world has been able to watch the 'walks' on the Moon. The Russians are much less publicity-conscious, and much of their work is veiled in secrecy so far as other countries are concerned.

Obviously it is difficult, therefore, for any detailed account of Soviet rocketry to be written at the present time; we must await an official history compiled by a Russian author. However, this book gives many of the data which have been made available. What is needed now, of course, is a complete union of the two programmes, so that future work can be carried out in peaceful collaboration. In time, this must come; in any case, historians of the coming centuries will look back with respect and admiration at all the space pioneers of our own time, no matter from what nation they come.

PATRICK MOORE

1 The Legend and the Record

Most people consider astronautics a heretical idea and refuse to entertain it at all. Others are skeptical, regarding it as an absolute impossibility while others are too credulous, considering it a simple matter easily accomplished. But the first inevitable failures will discourage and repel the fainthearted and destroy the confidence of the public.

K. E. Tsiolkovsky, 1929

THE STORY of early rocketry began before the birth of Christ, its exact origin lost in legend and antiquity. The recorded use of saltpeter (potassium nitrate) by primitive Eastern tribes for curing meat provides us with our first link in the evolution of propulsion and rocketry. The explosive use of this ancient household item was very likely discovered when some was dropped accidentally into a fire, resulting in a bright flash flame. These flame-supporting properties no doubt were responsible for man trying to combine saltpeter with wood for fire making. Since sawdust or fine wood particles were not available, the saltpeter was probably added to charcoal. Saltpeter and charcoal are two of the three ingredients of gunpowder. In this period, however, there is no record that sulfur, the third ingredient, was ever added. It is certain, however, that a composition containing saltpeter and charcoal, known as "Chinese Fire," was used in the East long before the time of Christ. The use of Chinese Fire for propulsion probably de-

1

veloped initially from its use in hollow bamboo rods or arrows. It must have been accidentally observed that the bamboo rods had a tendency to propel themselves due to the expansion of gases through the hollow tube.

The Aeolipile

Another important link in the evolution of rocketry was the demonstration in A.D. 160 of the aeolipile (named after Aeolus, the god of winds) by Hero of Alexandria, a Greek mathematician and scientist. This apparatus consisted of a hollow sphere mounted so that it could rotate between two supports that carried steam from a closed container suspended over a fire. The rotating sphere had two right-angle pipes located 180° apart on the sphere. The steam jet escaping through the pipes caused the sphere to revolve. The aeolipile was probably the first known device used to demonstrate the jet propulsion principle.

Earliest Rockets

The first reference to the rocket principle is recorded in a Chinese chronical, *T-hung-lian-kang-mu*, where the use of the reaction principle is dated A.D. 1232, during the Mongol siege of the city of Kai-fung-fu (Pien-king). During this siege, the Chinese used two new weapons. The first, called "heaven-shaking thunder," had a bomblike function and was dropped from the walls of the city on the invaders. The second, called "arrow of flying fire," is generally regarded as the first application of the rocket principle. The "arrow of flying fire" was probably an extension of the stuffed bamboo rod and most likely consisted of a small package of incendiary material tied to an arrow.

From this beginning, the use of gunpowder rockets grew rapidly, and the pyrotechnic art spread from Asia to the Middle East and then to Europe and England.

It is evident from chronicles written in the next few centuries that much progress was made in pyrotechnics and rocket design.

Roger Bacon, an English monk, is credited with the introduction of rockets into Europe prior to 1249. In "De Mirabili Potestate Artis et Naturae," he established the composition of gunpowder as follows: ". . . but of saltpeter take 7 parts, 5 of young hazel tweigs, and 5 of sulphur; and so thou wilt call up thunder and destruction, if thou know the art."

Early European Rockets

In 1258, the first mention of a rocket in Europe appeared in the Chronicle of Cologne, and, in 1379, an Italian historian credits a rocket with a significant victory in the battle for the Isle of Chiozza.

A German engineer, Konrad Kyeser von Eichstädt, experimented with gunpowder mixtures in 1405 and is credited with advancing the art of pyrotechnics.

A technical paper, entitled "Treatise upon Several Kinds of War Fireworks," published in 1561 in France, describes the use of military rockets in the defense of Orleans against the English in 1429, during the siege of Pont Andemer in 1449, against Bordeaux in 1452, and at Gand in 1453. This particular treatise is significant in that it suggests an alternate material for the rocket casings, which up to that time were still bamboo. This bamboo technique was probably the same one that had been in use for fifteen hundred years.

In 1630, the rocket evolved in the form of a grenade, and in 1645, during the Thirty Years' War, the rocket was credited with the downfall of Phillipsburg, France.

In 1668, Col. Friedrich von Geissler, a German, conducted the first purely scientific research in rocketry. Successful propellant experiments were conducted on a 55-pound and 132-pound rocket utilizing wooden rocket cases reinforced with linen.

Newton's Laws

In the latter part of the seventeenth century, Sir Isaac Newton opened the third period of rocketry when he interpreted and correlated many diverse observations and combined the results into three fundamental laws known as Newton's Laws of Motion. These laws simplified the science of mechanics and the Third Law forms the basis of modern rocketry. The Third Law states: "For every action there is an equal and opposite reaction, and the two are along the same straight line." Such a reaction takes place regardless of external conditions—underwater, in the air, or in a vacuum.

Sir William Congreve

In the accounts of the India Campaign in the latter part of the eighteenth century, the British forces in India were defeated with severe losses by an opposing Indian rocket force of 5,000 men. In this campaign, the rocket case had gone through still another step in its evolution. It was now made of iron, being some 8 inches long by 1.5 inches in diameter. However, the rocket still carried an 8-foot stick as a stabilizer.

It was at this time that Sir William Congreve became interested in the military rocket. His experiments at the Royal Laboratory at Woolwich resulted in several successful military rockets with ranges up to 2,000 yards.

During the European wars of the early nineteenth century, these rockets were successfully employed against Boulogne (1806), Copenhagen (1807), and Danzig (1813). Just about this time, two other uses for the rocket were found, namely bomb carrying and shrapnel.

By this time the military rocket had become so popular that, in 1817, the British Army formed the Field Rocket Brigade. The rocket brigade took part in virtually every important battle against Napoleon and distinguished itself particularly well in the final battle at Waterloo. Probably Congreve's rockets are best remembered for their use against Fort McHenry, which inspired Francis Scott Key to write, "The rockets' red glare, the bombs bursting in air," in "The Star-Spangled Banner."

The evolution and utilization of the Congreve rockets were milestones in the development of rocketry. They had first been developed as an incendiary rocket, which had limited military value; then the rocket was given bomb and shrapnel-type warheads resulting in a far greater destructive capability. Finally, rocket stabilization in flight, by the application of jet vanes in the nozzle, increased the rocket's accuracy and consequently its usefulness as a military weapon. Even with all these improvements, military rockets became obsolete at the end of the nineteenth century with the development of the more accurate artillery gun.

Early Russian Rocketry

Among the many documents dealing with the development of rocketry, we find numerous mentions of the introduction of rocketry into Russia. Some historians feel that the arrival of rockets in Russia coincided with the first use of gunpowder for military purposes. The rockets, however, were used solely for firework displays and celebrations and would be for many years to come.

In *Russian Rocketry, A Historical Survey*, I. A. Slukhai writes that rocket know-how and experience was handed down from generation to generation until the 1600s, when the accumulated experience was documented by the Russian gunsmith Onisim Mikhailov. Between 1607 and 1621, Mikhailov compiled the "Code of Military, Artillery, and Other Matters Pertaining to the Science of Warfare," at present the oldest preserved written document in existence in the Soviet Union. This document contains a detailed description of rockets, or, as the author calls them, "Cannon balls which run and burn."

V. N. Sokolsky, in his book *Russian Solid-Fuel Rockets*, however, disputes Russian rocket historians who based their information on Mikhailov's manuscript. These authors relied on the dates of the manuscript compiled by Mikhailov, between 1607–1621, but it was not printed until 1777–1781. Thus they regard its information on rockets as proof of their use before the beginning of the seventeenth century. One must remember that Mikhailov's manuscript is not an original work, but a collection

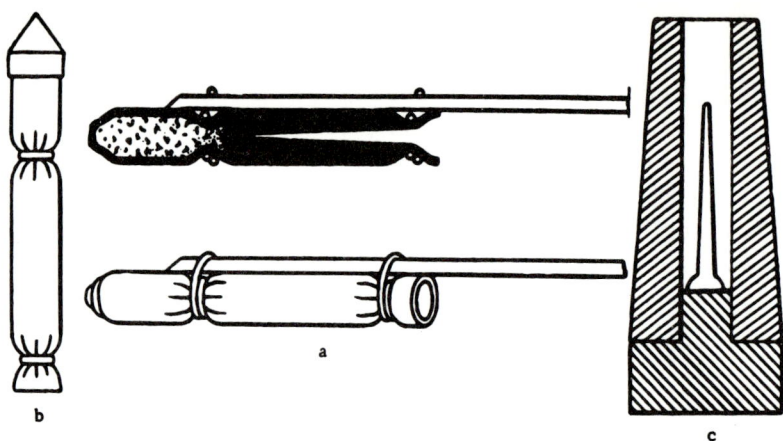

FIG. 1 Eighteenth-century pyrotechnic rockets: a) general view and cross section of rocket; b) rocket casing with cap; c) mold for packing rocket casings (hollow molding).

of 663 decrees or articles selected from foreign military books, therefore it cannot serve to confirm the use of rockets in Russia before the seventeenth century.

Sokolsky claims that the first reliable information on the use of rockets in Russia dates only from the second half of the seventeenth century. Specifically, he refers to a fireworks display that was held in the town of Ustyuga in 1675. He also provides additional proof to his claim by referring to a book by Balthasar Koiet that describes a birthday celebration for a member of the Tsar's family in 1675 which was marred by the accidental explosion of fireworks and rockets.

Peter the Great Unlike Sokolsky, who went to great pains to substantiate dates about Russian rocketry, other Russian historians were content to state that rockets were not used at all in Russia until the time of Peter the Great. Tsar Peter I devoted his life to the building of Russia's military might. One of the many things that he accomplished was, according to I. A. Slukhai, the founding in 1680 of the first Rocket Works in Moscow, where standard signal and illuminating rockets were made for the Russian army. Later in the 1700s, Peter moved the Rocket Works to his new capital in St. Petersburg and vastly

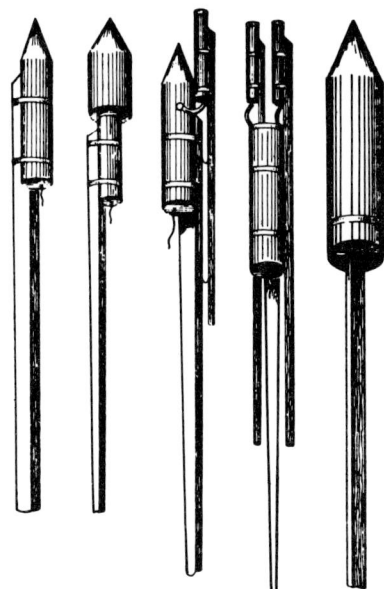

FIG. 2 Pyrotechnic rockets at the beginning of the nineteenth century.

expanded its operation. Peter hired English, Scottish, Dutch, German, and French officers to modernize his armed forces. His Russian soldiers fired rockets under the guidance of this Western command. The Scotsman, Patrick Gordon, who was both a general and an admiral in Peter's service, mentions in his diary Peter's personal supervision of the production of rockets for his entertainment. The primary use of rockets in Russia was still, at this time, limited to celebrations and signaling.

Alexander Zasyadko   Alexander D. Zasyadko (1779–1837), an officer of the Tsar's artillery, after noting Congreve's inventions and studying the Rocket Works' files accumulated in the 135 years of the factory's existence, came up with rocket designs of his own. Some of these rockets were tested in 1817 in St. Petersburg and were successful. As a direct result of these successes, Zasyadko was assigned to western Russia to train the Tsar's first soldiers to use military warhead rockets. In the following year, Zasyadko was promoted to major general and made head of Russia's first artillery school. Beginning in the

7

1820s, the production of rockets in Russia was concentrated first at the St. Petersburg Pyrotechnic Laboratory and subsequently at the St. Petersburg Rocket Institute (P.R.Z.) established expressly for that purpose.

Zasyadko's solid fuel rockets were first used in the Russo–Turkish war of 1828–1829, during the sieges of Varna, Schumla, Silistra, and Braila. Also, Russian ships armed with the rockets operated successfully in the Black Sea and on the Danube. Zasyadko emerged from the war as a national hero. Later after his death, his rockets, by the thousands, were deployed in the conquest of the Caucasus and more widely in the Crimean War of 1853–1856. In 1959, at the Exhibition of Achievements, Nikita Khrushchev added the portrait of General Zasyadko to the collection of the Moscow Space Museum.

Konstantin Konstantinov Zasyadko was followed by Konstantin I. Konstantinov (1817–1871), also an artillery officer with still a greater mathematical capability than his predeces-

FIG. 3 *Left:* High-flying incendiary rocket.
Right: Rebounding rocket with explosives, designed by Zasyadko.

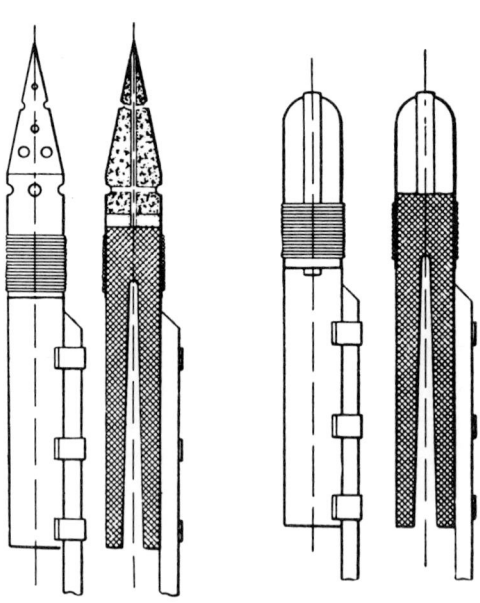

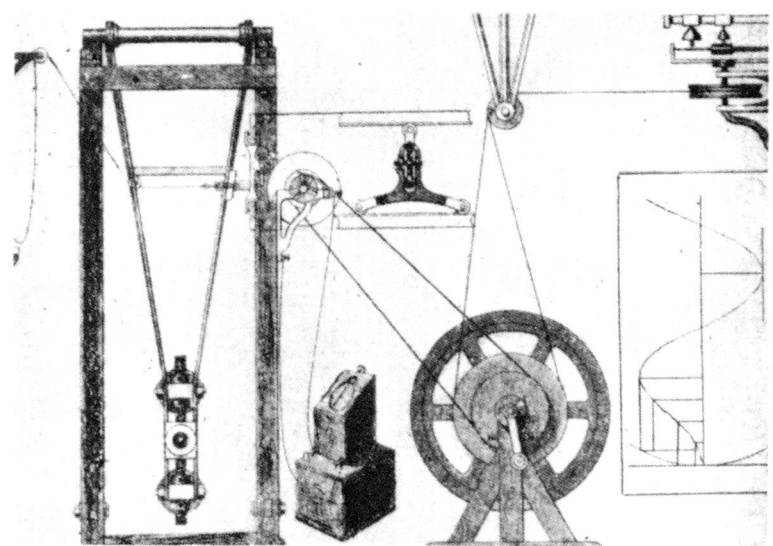

FIG. 4   Konstantinov's rocket ballistic pendulum.

sor. The Russians called Konstantinov the founder of experimental rocket dynamics and the first to produce gunpowder rockets on a mass scale. In 1847, at the age of thirty, he started working on practical problems of rocket production. Up to that time, theory was nonexistent and the production of rockets depended on experiment and the skill of individual experts. Konstantinov appreciated the fact that for rocket development to continue, it was necessary to create "a mathematical theory of the construction and firing of rockets." To this end, he developed a ballistic rocket pendulum to measure the relative force of rockets, both at maximum thrust and its variation at the beginning and at the end of combustion. In 1850, he was appointed commandant of the St. Petersburg Rocket Institute and within several years remodeled it almost entirely.

Like Zasyadko, Konstantinov's first task was directing large-scale rocket production, but he also spent a great deal of time training rocketeers, organizing rocket batteries, and working out the tactics of rocket operations. In addition to military application, he also studied nonmilitary uses of rockets. In 1882, he successfully designed and tested a rescue rocket for shooting lifelines to shipwrecked vessels. Later he suggested the use of the rocket as a travel vehicle.

9

I. I. Tretesky   In the second half of the nineteenth century, in Russia as in other European countries rockets lost their military importance because of the high state of development of the artillery cannon. But at the same time as the popularity of the rocket as a war weapon decreased, interest in the rocket as a propulsive source in aeronautics and aviation increased. In March, 1849, a Russian artillery captain and military engineer, I. I. Tretesky (1821–1895) proposed in his paper, "On Methods of Guiding the Aerostats," using gunpowder, steam, gas, or compressed gas to propel a dirigible. The idea was later rejected by the Military Staff College as impracticable.

FIG. 5   Russian two-inch military
rockets (1859–1867).

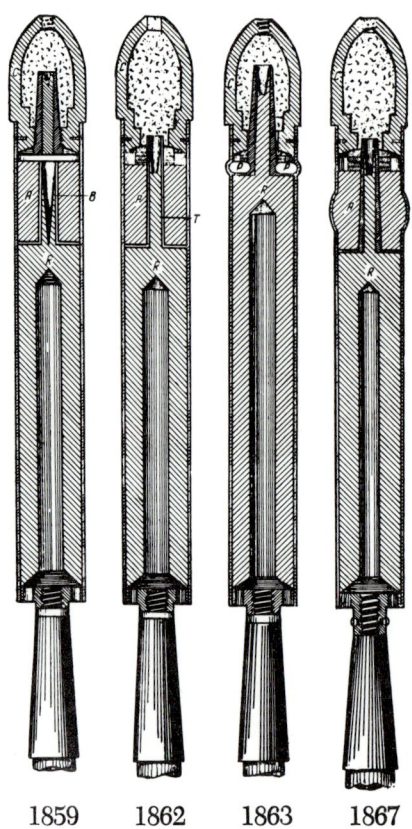

1859   1862   1863   1867

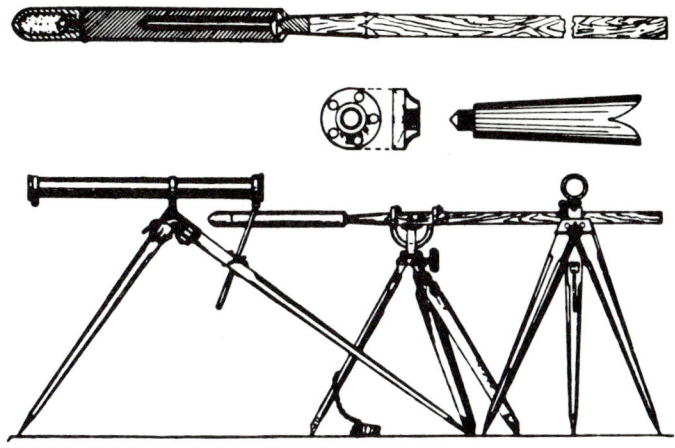

FIG. 6   Details of two-inch rocket and general view of rocket launcher.

The use of compressed air for propelling a dirigible again found a proponent in Admiral Nicholas M. Sokovnin (1811–1894). In 1866, the admiral, in his book *The Airship,* wrote that he always had been convinced that an airship must fly in the same manner as a rocket. Accordingly, he suggested a turtle-shaped dirigible moved by the rocket principle with the aid of compressed air released from pipes located on both ends.

In 1870, Tretesky once again evolved a plan for utilizing jet propulsion to propel dirigibles. In this concept he proposed using gunpowder. But as before, his idea was rejected.

Nikolai Kibalchich   Up to this period in time, the application of jet propulsion was visualized as propelling lighter-than-air flying machines, essentially balloons. In 1881, Nikolai I. Kibalchich (1853–1881) is credited with first proposing the idea of heavier-than-air machines rising vertically with the aid of rocket propulsion and carrying human pilots and passengers. Kibalchich was the bomb expert of a revolutionary organization called *Narodnaya Volya* (People's Will). After a number of unsuccessful attempts on the life of Tsar Alexander II, a bomb prepared by Kibalchich was thrown at Alexander, mortally wounding him. Kibalchich was seized by the Tsar's police on March 17, 1881, and sentenced to be executed shortly thereafter.

11

FIG. 7   Nikolai Ivanovich Kibalchich, 1853–1881.

Kibalchich's rocket has been described as a platform with two upright posts. The rocket motor was mounted between these posts in a manner that would permit the motor to be swiveled from the vertical position (to provide an upward thrust) to the horizontal position (to provide a forward thrust). One of the more unique features of the concept was the special mechanism that feeds gunpowder cartridges to the motor's combustion chamber. The propellant feed was to be continuous and automatic. Although before his death he asked that his notes be given to scientists for study, his request was denied and his project was buried in secret archives.

Thirty-six years later, in August, 1917, his notes were found and published for the first time. The preface to the article was written by N. A. Rynin, a prominent Russian author on rocketry and space flights who said that Kibalchich's name should be established as the first to advance the idea of the use of jet engines for flying. Tsiolkovsky went on to develop the same idea and acknowledged his indebtedness.

Jet Flight

Interest in jet flight continued to occupy the minds of the imaginative throughout the remainder of the nineteenth century. In 1886, A. V. Evald proposed a jet propelled airplane with rocket engines, using gunpowder arranged in a special chute. After several unsuccessful experiments he eventually achieved success, but, as with most inventions, no additional tests were made because of the lack of funds.

F. Geshvend   Another was F. Geshvend's plans for a jet-propelled heavier-than-air flying machine, the steam-jet airplane. In 1887, Geshvend designed a biplane with a highly complicated body and wing structure. The unique feature of this design was the use of nozzles emitting a steam jet that would suck in a greater mass of air, augmenting the thrust. In the same year he introduced an important change in design by eliminating the use of wings and replacing it with vertical jet streams. He retained small wings (one-fifth the size of the first design) for stabilization and indicated that his steamplane could move

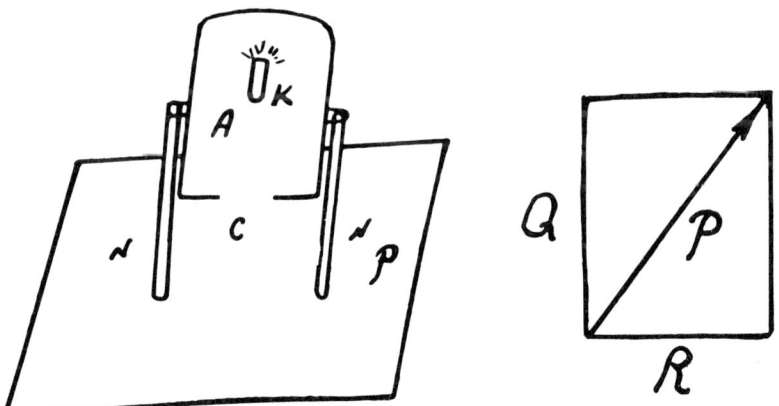

FIG. 8   Schematic diagram of Kibalchich's rocket.

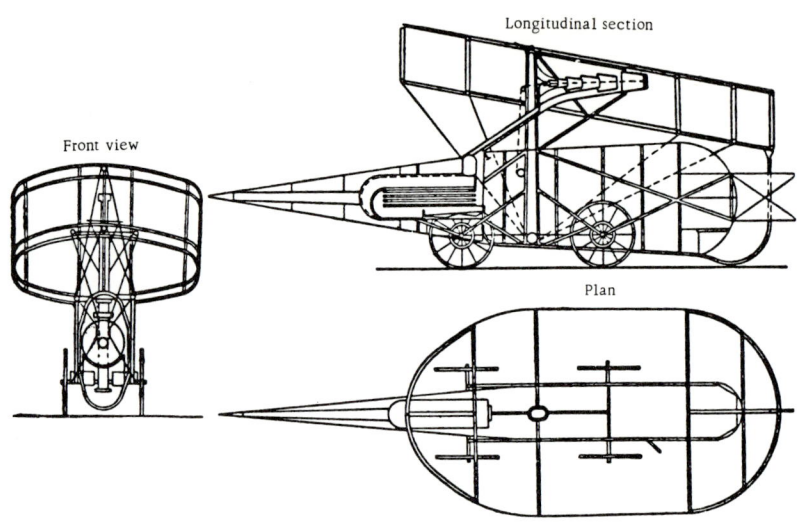

Longitudinal section

Front view

Plan

FIG. 9   Geshvend's "steamplane."

in any direction and hover in place. "By operating one vertical jet," he wrote "it is possible to remain in one place in the air without movement for more than one hour, the hourly steam and water expenditure being less than the tank capacity. . . ." Geshvend's new design was important in the chronology of rocket development in that, unlike his original design but like Kibalchich's, it was based on the rocket principle, i.e., it did not require the atmosphere as a supporting medium.

In 1889, Geshvend's first design was examined by a Military Science Commission and rejected, because it had not been tested and was based on the airplane principle, which the commission did not consider promising. Needless to say, the commission entirely overlooked his second design, which was predicated on the rocket principle.

A. P. Fedorov   In the late 1880s, such people as V. D. Spitsyn, S. M. Nemirovsky, P. N. Lebedev, and A. P. Fedorov all worked on jet propelled flying machines. Fedorov's design is the most interesting of those produced during this period. His machine was to be propelled by means of a system of tubes. Fedorov wrote:

14

Our tube, just as a rocket in flight, or a recoiling rifle, is impelled to move along its own axis . . . by a thrust . . . directed along its longitudinal axis from the open end to the closed end. . . . If we construct a system of tubes in which (1) one group stands vertically, with the exhaust downwards, (2) others lie horizontally along the longitudinal axis of the machine, and (3) still others are coiled helically around the vertical axis, the first group will produce lift, the second translational movement, and the third, rotation about the vertical axis acting as a rudder; consequently our system possess everything required for free flight.

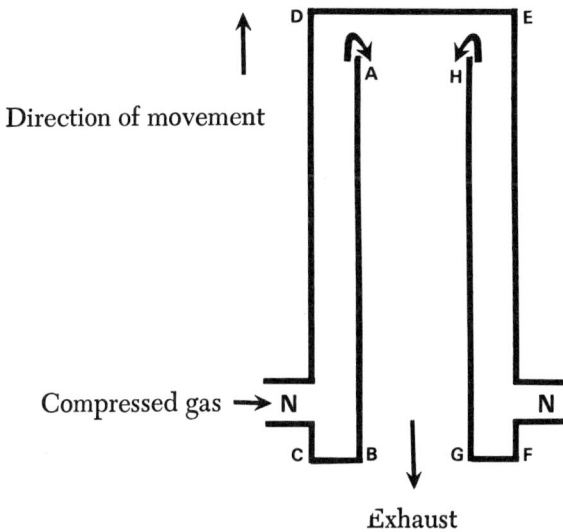

FIG. 10   Cross section of the tubes (engine) of Fedorov's airship. Compressed gas enters at N and fills inner cylinder. The gasses then pass through opening A H into chamber A,B,G,H, exhausting through opening B-G.

The next phase of Russia's rocket development aimed toward interplanetary travel and is inseparably linked with Konstantin E. Tsiolkovsky (1857–1935), the founder of astronautics in Russia.

FIG. 11   Konstantin Eduardovich Tsiolkovsky, 1857–1935.

2 Tsiolkovsky's Legacy

For a very long time, like everyone else, I regarded the rocket as a little more than a toy with a few more or less trivial applications. . . . I do not remember how I got the idea of making computations relating to rockets. . . . It seems to me that the first seeds of thought must have been sown by the fantastic novels of the famous Jules Verne; they set my brain working in a definite direction. I mused; musing led me to more serious mental activity. Of course, it would never have come to anything if there had been no support forthcoming from the side of science.

K. E. Tsiolkovsky, 1911

KONSTANTIN EDUARDOVICH TSIOLKOVSKY (1857–1935) is considered by the Soviets to be the "Father of Soviet Space Flight." His work and his life story have become a legend in his own country. But like the stories of most national heroes, after a time, it becomes difficult to separate the man from the myth. Fact or fiction, in the Soviet Union Tsiolkovsky's life and legacy is a source of inspiration to men of science and to young people aspiring to the scientific way of life.

Tsiolkovsky was born in 1857 near Moscow, the son of a forestry expert and amateur inventor. His early boyhood was a normal one until the age of nine, when he contracted scarlet

17

fever and through a series of complications became almost totally deaf. His deafness kept him from school, and he taught himself from books that were in his father's meager collection and from others that he could borrow. He mastered mathematics first and then physics. His father noted with awe his dedication to learning and decided that no matter what the sacrifice, Tsiolkovsky was to attend Technical School in Moscow. He remained there for three years until he was forced to return home, where he helped support his family by tutoring. At twenty-two, in spite of the lack of a formal education, he passed the required examinations, and was granted a certificate as a "peoples' school teacher," the lowest rank in the Tsarist educational system. Shortly after he married, and in 1881 he began his first serious scientific research, in three areas: the development of an all-metal dirigible; an airplane; and a rocket for interplanetary travel. His approach to all of these problems was as an inventor rather than a builder.

Metal Dirigibles

In 1883, ten years after Tsiolkovsky conceived the idea of conquering space by means of centrifugal force, he came to the conclusion that only by the principle of jet propulsion could space flight be achieved. In his manuscript, "Free Space," he stated that space travel, in the absence of gravity and air resistance, had to be based on the reaction of particles being ejected from a body traveling in space and that motion was impossible without the loss of matter.

It was during the period of 1885 through 1892 that Tsiolkovsky carried out most of his work on metal airships. He designed his dirigible with movable sides so that he could vary the volume of the dirigible and in that way maintain a constant lift at varying air temperatures and altitudes. Also, the internal volume of the gas could be heated and expanded by having the dirigible's gas coils placed by the engine's exhaust.

Tsiolkovsky first submitted his idea to the Russian Engineering Society and the plan was rejected on the basis that dirigibles "will eternally be the playthings of the wind." He then sub-

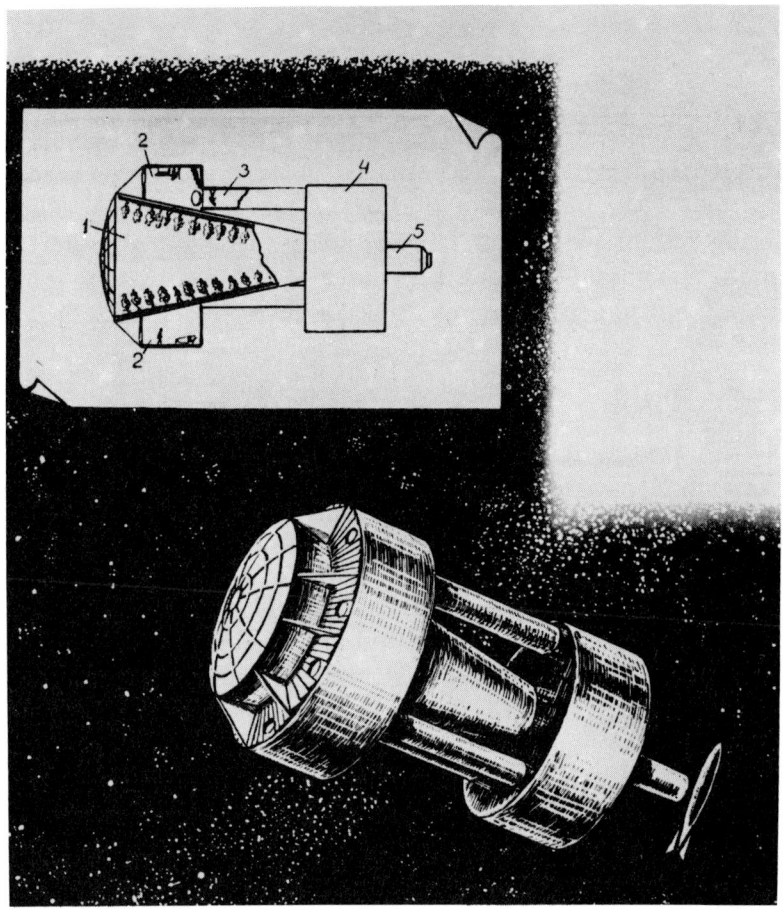

FIG. 12   Space station as visualized by Tsiolkovsky: 1) closed ecological system and garden; 2) laboratory area; 3) passage to storage and living areas; 4) storage and living areas; 5) docking mechanism.

mitted his idea to the General Staff of the Russian Army and again the concept was rejected. Before discontinuing his investigations of dirigibles in 1892, he published a paper entitled "Metal Dirigibles." The only response was a few sympathetic reviews.

Interplanetary Flight

Although little recognition was being given to Tsiolkovsky's theories and proposals, his creativity did not diminish. He continued to develop his ideas about interplanetary flight in his science fiction stories, *On the Moon* (1893) and *Dreams of Earth and Heaven* (1895). In the latter story, he presented his first ideas on the creation of an artificial earth satellite. Tsiolkovsky wrote about "an imaginary earth satellite like the moon, but brought arbitrarily closer to our planet, to a point barely outside the limits of its atmosphere, that is, about 200 miles from its surface."

Other Inventions

Tsiolkovsky is also credited with proposing the first all-metal airplane. In an article written in 1894 entitled, "The Airplane, A Birdlike Flying Machine," he gave a description and provided sketches of a monoplane. The plane had wings with rounded edges and a streamlined fuselage. The external configurations were comparable to the structures that were developed by aircraft engineers about fifteen years later.

Tsiolkovsky built the first wind tunnel in Russia in 1890, carrying out more than 1,000 experiments in aerodynamics. It was during this same period that he recognized the value of the internal combustion gasoline engine as a source of power for aircraft. However, his airplane, like the metal dirigible, was rejected by official Russian scientific circles. With no support and lacking funds, he finally abandoned this area of research.

Rocketry

He then turned to rocketry and as early as 1897 put together his well-known formula establishing the relationship between the velocity of the rocket jet and the velocity of the rocket, and the mass of the rocket and the mass of the propellant burned for power. Initially, Tsiolkovsky's formula gave the rockets ideal velocity without taking into consideration flight losses due to drag and gravity. Subsequently, he developed a solution that took these factors into consideration, providing a method for calculating the flight of a missile more realistically.

Space Flight

In 1903, Tsiolkovsky published his classic paper, "Investigating Space with Reaction Devices," in the magazine *Survey of Science*. In the paper, he elaborated on his formula, which made it possible to determine the flight performance of a rocket if the propellants and propellant weights are known, and he proved that it was possible to attain orbital and escape speeds. His paper was not confined to theoretical calculations. He also gave practical instructions on rocket design and examined a great many rocket propellants. But Tsiolkovsky's ideas were so far ahead of their time that during the first decade of the twentieth century his work went unnoticed. In 1911, his paper about space flight was printed a second time. An adaptation of the original article, it was a nontechnical version of his ideas on space flight and received a wider distribution. In this particular article, he stated that some new form of energy, far surpassing liquid propellants, had to be found for space flight and proposed the use of atomic power for space flight—meaning, in his terms, using the energy of atomic decay.

A year after the Russian Socialist Academy was organized in 1918, Tsiolkovsky was recognized for his work by being elected to the Academy, and a personal pension was granted him by the Commission for Improvement of the Lot of Scientists (TsEKUBU). In 1923, the Socialist Academy was renamed the Communist Academy, and, in 1936, the principal institutes of the Communist Academy were tranferred to the Academy of Sciences U.S.S.R.

In the next few years, Tsiolkovsky devoted a great deal of his time to the difficult problem, even by present-day standards, of returning from space and landing on the earth. He also struggled with theories involving the potential use of light pressure for space flight and other exotic propulsion schemes, such as the use of electromagnetic waves.

In 1925 and 1926, his work carried him briefly back to the idea of building nuclear and electrojet engines, but his primary preoccupation seems to have been centered around ways and means of attaining the high velocities essential to attaining interplanetary flight.

In 1926, he investigated the use of an automobile, steamship, locomotive, airplane, dirigible, gas and electromagnetic cannon as an auxiliary means of providing an initial high boost velocity. His studies showed that at best he could only obtain about 300 to 650 ft per sec, which led him to the conclusion

> that to give the projectile a velocity of over 650 ft per sec special means are required. . . . In this case, the simplest and cheapest way is to use a rocket or jet: that is, our space vehicle must be placed in or on another land rocket, which, without itself leaving the ground, gives the desired take-off speed.

In 1924, Tsiolkovsky discussed the subject of multistaged rockets, in his book *Cosmic Rocket Trains.* He suggested the use of a two-stage rocket in which the first stage or the "land rocket" would move on the earth and in the lower and denser layers of the atmosphere, and the second stage would go on by itself to achieve velocities compatible with interplanetary flight. The lower stage would return to earth after expending its propellant supply.

In the last years of his life, he worked on the development of a theory for jet airplane flight. In his article "Jet Airplanes" (1930), he wrote in detail about the advantages and disadvantages of jet airplanes in comparison with propeller aircraft and concluded the article with the statement, "The era of the propeller airplane must be followed by an era of jet or stratosphere airplanes."

In 1935, in another paper, "The Maximum Velocity of a Rocket," Tsiolkovsky once again returned to the problem of

FIG. 13   Straight-nozzle rocket proposed by Tsiolkovsky in 1903.

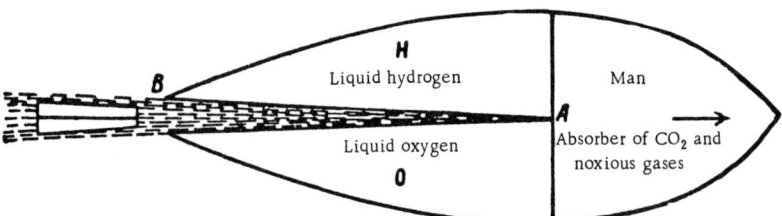

multistaged rockets, believing them to be possible and practical. He also realized that technically they were beyond all existing capabilities at that time. He then proceeded with a disclosure of a simplified version of a multistage rocket, which he referred to as a "rocket squadron." Unlike the rocket train (one rocket on top of the other), the rocket squadron was a parallel configuration (several rockets connected in line with one another). Other different points in the squadron design was that all of the rockets in the squadron would ignite their engines simultaneously at lift-off. But, instead of consuming all of their propellants, they would consume only half, at which time they would transfer the remaining propellants into the adjacent rockets, which also would be half empty. The empty rockets would then be jettisoned and return to earth, and the now full rockets would continue. This process would be repeated until one rocket would finally achieve orbital or interplanetary speeds.

Soviet Hero

In recalling his life, Tsiolkovsky compared his struggle for recognition with the struggles of other visionaries and jotted down his thoughts in the foreword to his paper, "Rocket to Outer Space":

> Lamarck wrote a book in which he analyzed and demonstrated the development of creatures from the lowest organism to man. The French Academy, with the renowned Cuvier as its head, derided the book and publicly called Lamarck a donkey. Galileo was tried and imprisoned and forced ignominiously to retract his teachings about the earth's rotation. Only by doing so was he saved from the stake. Kepler was imprisoned. Bruno was burned for teaching that there is a multiplicity of worlds. The French Academy rejected Darwin, and the Russian Academy, Mendeliev. Columbus, after discovering America, was put into chains. The derision of scientists led Mayer to the madhouse. The chemist Lavorser was put to death. . . . There is no counting those that have been burned and hanged for the sake of truth. History is full of such things. And why have academies, scientists, and pro-

fessionals been condemned to play such a wretched role as extinguishers and even chastisers of truth?

Six days before his death in September, 1935, Tsiolkovsky wrote a letter to the Central Committee of the Communist Party, bequeathing all of his works on aviation, rocketry, and interplanetary travel "to the Bolshevik Party and the Soviet Government." His library manuscripts, models, and other memorabilia were taken to the central offices of Aeroflot, but later put into several museums.

Tsiolkovsky's house in Kaluga is now a museum and contains some of his books and personal belongings. During World War II, the Nazis occupied Kaluga and stole or destroyed some of the objects on display. However, most of the exhibits were saved, restored, and replaced after the war. The scientist's house is now a national Soviet shrine and tourist attraction. In 1952, a large gold medal in his honor was minted by the Aero-club of France. Two years later in 1954, the Soviet government established the Tsiolkovsky Gold Medal, which is to be awarded every third year to the most outstanding contributor to space flight.

In 1957, a full-length motion picture was made about Tsiolkovsky's life entitled, "The Road to the Stars." A monument was erected to him in Kaluga in 1958 in the form of a silvery, upright model of a rocket, approximately 60 feet high, serving as the background for Tsiolkovsky's bronze figure. The granite pedestal at the base of the figure is inscribed K. E. TSIOLKOVSKY, 1857–1935, and includes the prophecy from a letter written by him to B. N. Vorobyev on August 12, 1911:

MANKIND WILL NOT REMAIN FOREVER CONFINED
TO THE EARTH. IN PURSUIT OF LIGHT AND SPACE
IT WILL, TIMIDLY AT FIRST, PROBE THE LIMITS
OF THE ATMOSPHERE AND LATER EXTEND ITS CONTROL
TO THE ENTIRE SOLAR SYSTEM.

3 Interplanetary Communications

Instead of cannons or aerostats, I propose the use of reaction machines to explore the atmosphere. By reaction machine, I mean a kind of rocket, but a specially designed rocket on a grandiose scale. The idea is not new, but the calculations yield such remarkable results that they simply cannot be ignored.

K. E. Tsiolkovsky, 1903

HISTORICALLY, Tsiolkovsky's ideas on "interplanetary communications" (the Soviets' terminology for space travel) can best be classified as the conceptual phase of rocket development in the Soviet Union. He is credited with providing the first seeds of thought that served as a scientific point of departure for some of his contemporaries and future Soviet scientists.

Space Scientists

In order to transform Tsiolkovsky's ideas of space travel into reality, Soviet scientists had to solve many complex problems in the fields of rocket theory, engineering, flight dynamics, and communications. In fact, Tsiolkovsky's goals required about forty years of growth and evolution and the contributions of several generations of rocket scientists. Then and only then did the Soviets bridge the concept/technological gap and launch their first earth satellite. On its successful attainment, Academician A. A. Blagonravov, of the Soviet Academy of Sciences, said:

. . . Space flight is not the accomplishment of a single man. It involves creative search and the strenuous effort of thousands of people, it embodies lofty attainments in many fields of . . . science and engineering. . . .

The pioneering few among the thousands who are recognized for their specific contributions, include, beside Tsiolkovsky, Ivan V. Meshchersky, Nikolai E. Zhukovsky, Fridrikh A. Tsander, and Yuri V. Kondratyuk.

Reaction Principle

Tsiolkovsky's chief contribution lay in his conclusion that the only possible means of flight in space must be based on the reaction principle and that a rocket accelerates in powered flight because it is constantly losing weight as the propellant is burned. Thus, the flight characteristics of a rocket could not be solved by the equations of classical mechanics, founded on Newton's three laws, but instead required a new branch of theoretical mechanics, dealing with bodies of variable mass.

Ivan Meshchersky   A primary role in the solution of that problem was played by Ivan V. Meshchersky (1859–1935), a professor at the St. Petersburg Polytechnic Institute. For his doctoral thesis (1897–1904), Meshchersky prepared and published his now classical work, "The Dynamics of a Point of Variable Mass." In this paper, Meshchersky showed that "in our study of the motion of bodies of variable mass, we come upon the problem of the motion of material points whose mass changes in the course of time." He took as an example the vertical flight of a rocket:

> As the rocket flies upward its mass decreases because of the combustion of the fuel which propels it; the main forces acting on the rocket are the forces of gravity, the air resistance, the pressure of gases evolved by the burning propellants and an additional force due to the burnt particles being ejected with some relative velocity.

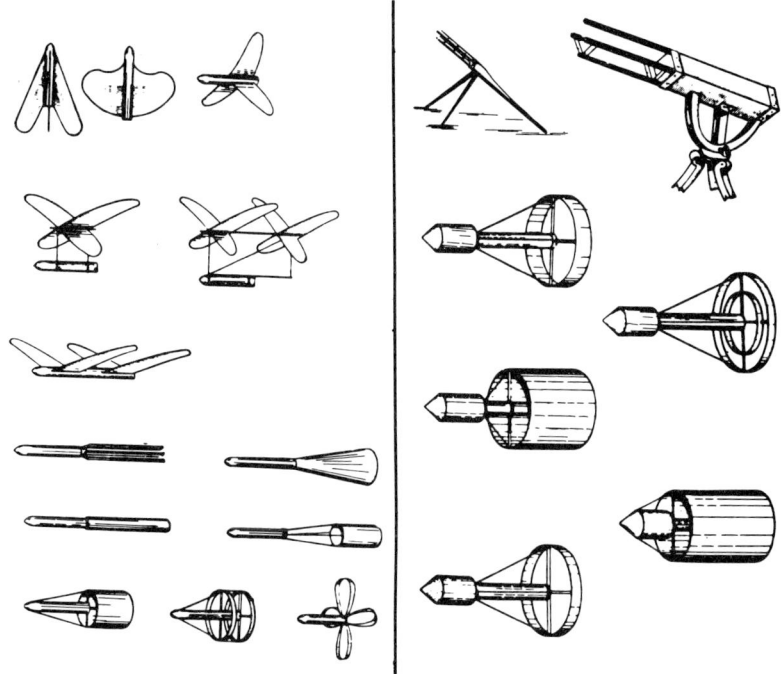

FIG. 14   *Left:* Pomortsev's sketches of experimental rocket designs with different types of stabilizers. *Right:* Rocket designs with launching stands.

The publication of his thesis came at a particularly difficult period in the Soviet Union. At the end of the nineteenth century, rockets had no specific practical application. His work did not go by unnoticed, however, for, in 1959, he received top recognition for his work in theoretical mechanics: Pravda, the official Soviet newspaper observed the 100th anniversary of his birth by issuing a headlined article stating that, "It Is to His [Meshchersky's] Labors that Sputnik Owes Its Existence."

Other Researches

Despite the lack of a practical need of rockets in the early 1900s, progress continued along many fronts in the Soviet Union in the fields of scientific education and investigation. By 1902, a wind tunnel was built and installed at the Moscow Uni-

versity by Prof. Nikolai E. Zhukovsky, the "Father of Russian
Aviation." (Tsiolkovsky had built the first one.) During the
years 1904 and 1906, Dmitri P. Ryabouchinsky designed and
built the famous Kuchino Institute of Aerodynamics; and still
later, at the close of the first decade (1910–1912), Zhukovsky
completed still another aerodynamic laboratory in the Moscow
Higher Technological College.

Also, at this time (1911–1912) Tsiolkovsky published his
expanded version of his famous 1903 paper, "The Investigation
of Space with Jet Machines." Then, in 1914, he published a
supplementary pamphlet to the papers in which he pointed
out the possible use of ozone as a propellant.

Throughout this period Professor Zhukovsky lectured on
general aerodynamics and Prof. S. A. Chaplygin (1869–1942)
conducted basic research in the area of theoretical gas dynamics. Ryabouchinsky, the builder of the Kuchino Institute,
working with Gen. Mikhail M. Pomortsev designed and tested
a rocket bazooka. And, still other Russian scientists, including
Prof. Nickolas A. Rynin, began writing and publishing popular
books and articles on rocketry and space flight.

Science Education

Thus, just prior to World War I, Russia had its share of
well-known aerospace pioneers, academic institutions, laboratories, and a cadre of space scientists in training, who were
to become established in their own time. Despite talent and
the institutions, however, the turning point in Russian rocketry did not occur until after World War I and the Revolution in Russia. As early as 1918, Lenin authorized Professor
Zhukovsky to build and staff a modern center of aeronautical
research, later named the *Tsentralyni Aero-Gidrodinamichescky
Institute* (Ts.A.G.I.). This organization was equivalent to
the U.S. National Advisory Committee on Aeronautics
(N.A.C.A.), later to be replaced by the *National Aeronautics
and Space Administration* (NASA) during President Eisen-
hower's administration. Ts.A.G.I. was partially completed in
1924 and became fully operational by 1927.

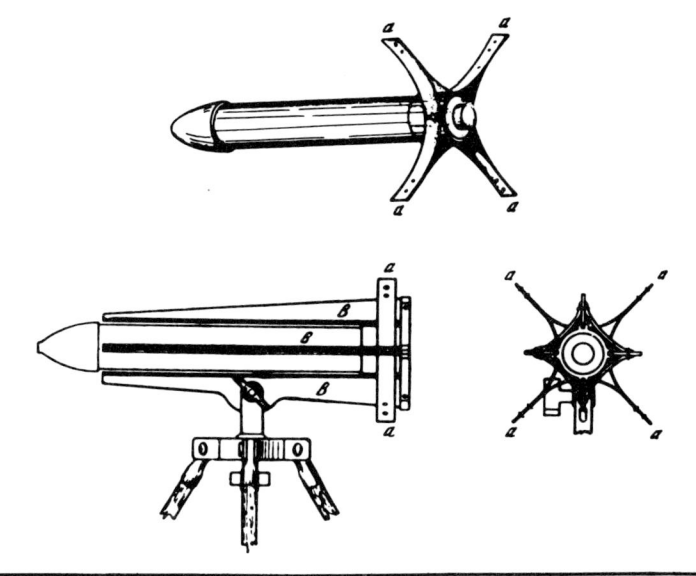

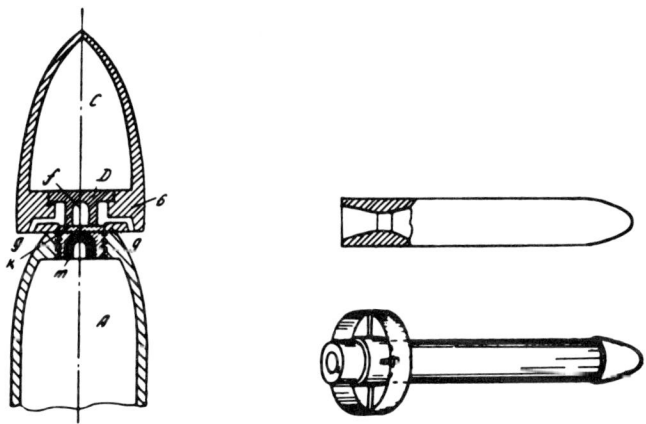

FIG. 15 *Top:* Rocket with star-shaped (fin) stabilizer and launching stand designed by Pomortsev (1902–1907). *Bottom:* Plan of Pomortsev's pneumatic rocket with annular stabilizer.

Also, in 1919, the famous Zhukovsky Academy of Astronautics, with its advanced teaching, research, and experimental facilities, was founded. Since then, at least fifteen comparable academies of astronautics have been built, and today a still larger number of technical colleges and universities offer courses in the aerospace sciences.

All of these institutions have not only been responsible for developing future aerospace scientists but have had the added responsibility for conducting basic national research. In this manner, the training of engineering talent and the flow of new technology into the aerospace field is assured.

Research Development Plans

In consonance with the academic activity there were a number of five-year industrialization plans that helped to establish and support the modern expanding aerospace establishment. The aerospace industry, by the snowball effect, created the proper environment for the development of basic research organizations. One of these was the newer Ts.A.G.I., built in 1933–1938 with more modern wind tunnels and other research facilities; another was the Ts.A.G.I., built in 1959–1960, which housed the Institute of Fluidmechanics of the Academy of Sciences in Novosibirsk, and the most recent was the installation of the Mach 8 wind tunnel at the Institute of Theoretical and Applied Mechanics of the Novosibirsk Research Center.

Additional technical support came from such organizations as the following: the Scientific Institute of Air Instruments (N.I.I.A.P.); the Flight Research Institute (L.I.I.); Central Institute of Aero Space Materials (V.I.A.M.); the Scientific Rocket Research Institute Number One (RNII.); the Scientific Testing Institute of the Soviet Air Forces (N.I.I.V.V.S.); and others.

Interplanetary Travel

As a result of the work of such men as Russia's Tsiolkovsky and Tsander, America's Goddard, and Germany's Oberth, the idea of interplanetary travel continued to spread within the

Soviet Union. Support and growth of the infant aerospace activities continued throughout the 1920 decade.

In 1924, the government created and supported a special
Central Bureau for the Study of the Problems of Rockets
(Ts.B.I.R.P.). This organization had the following objectives:

1. To bring together all persons in the Soviet Union working
on the problem.
2. To obtain as soon as possible full information on the progress
made in the West.
3. To disseminate and publish correct information about the
current position of interplanetary travel.
4. To engage in independent research and to study in particular
the military applications of rockets.

In the same year, 1924, the All-Union Society to Study Interplanetary Communications (O.I.M.S.) was established in Moscow. This group's objectives were comparable to Ts.B.I.R.P.'s.
The society had 150 members and was divided into three sections: research, popularization, and literary propaganda. Among
the members of the society were V. P. Vetchinkin, Y. I. Perelman, F. A. Tsander, K. E. Tsiolkovsky, and other engineers,
scientists, and public figures.

The Soviet space program had its origins in these early rocket
societies. The probable evolution of these groups which subsequently were responsible for the Vostoks and the sputniks is
traced chronologically in the chart that follows, which not only
depicts the organizations but also some of the key personnel.

Not depicted on the chart is the support of the space pro-
gram provided by the military, academic institutions and the
U.S.S.R. Academy of Sciences. Space objectives and priorities,
however, are established at the highest government levels, and
implementation is carried out through such organizations as the
State Commission for Space Exploration (S.C.S.E.). The
S.C.S.E. is thought to be concerned with the design phase of
space exploration planning. The Commission on Interplanetary
Communications (I.C.I.C.) was organized in 1957 and had
L. I. Sedov as its first Chairman and M. K. Tikhonravov as its
Vice Chairman. In 1963, the I.C.I.C. was renamed the Com-
mission for the Exploration and Utilization of Space (C.E.U.S.).

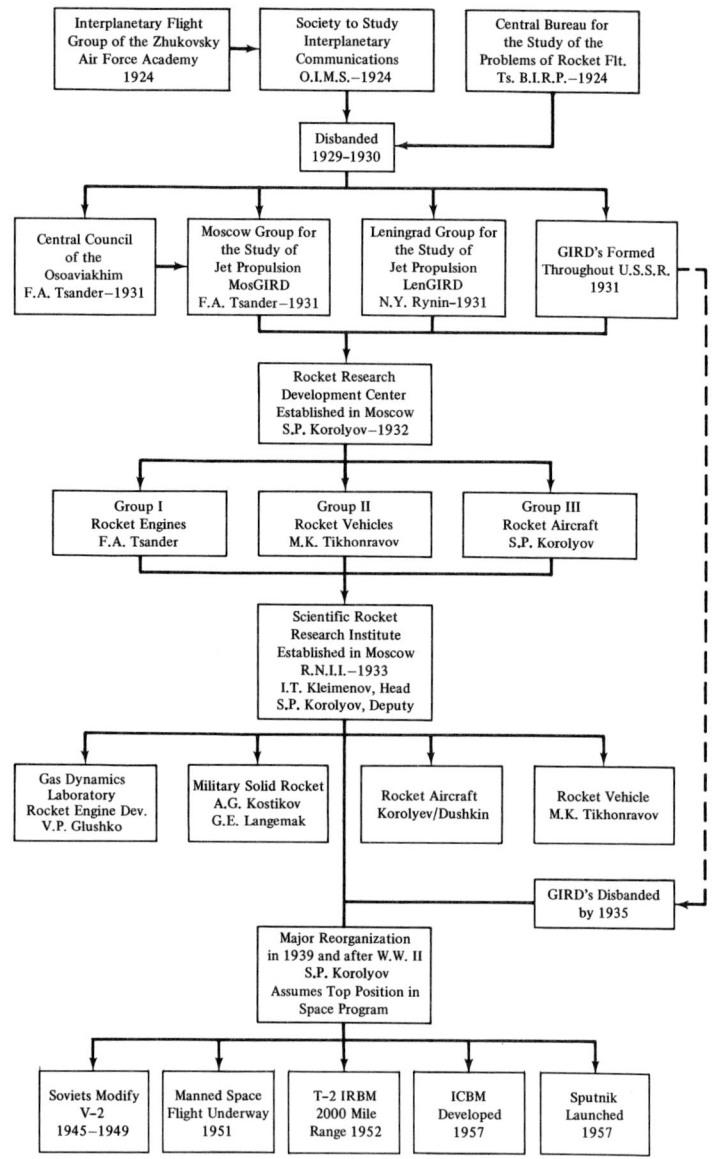

CHART I. Evolution of Soviet Rocket Societies

Its first chairman was Lt. General A. A. Blagonravov, a full member of the Academy of Sciences.

Chronologically, O.I.M.S. and Ts.B.I.R.P. were founded three years before the German Society for Space Travel and six years before the American Interplanetary Society (the name was later changed to the American Rocket Society and still later to the American Institute of Aeronautics and Astronautics). By 1927, Ts.B.I.R.P. and O.I.M.S. organized the first Soviet International Exhibition of Rocket Technology which was held in Moscow.

Boris Stechkin   In 1929, a new name, Boris S. Stechkin, was added to the growing list of outstanding researchers emerging in the Soviet Union. Stechkin, a pupil of the great Zhukovsky, was an ordinary engineer at Ts.A.G.I. when he published his work on the theory of jet propulsion, entitled "The Theory of the Air-Breathing Engine." The paper served as the theoretical foundation in the development of all types of air-breathing engines, including ramjets.

During the years 1928–1930, numerous rocket conferences were held, some closed, others open to the public. The major theme of these conferences was "the next step": what should be done in the Soviet Union in order to progress from purely theoretical rocket work to a more comprehensive and practical rocket program.

Two men who were especially active during this period were Fridrikh A. Tsander (1887–1933), one of the first disciples of Tsiolkovsky, and Yuri V. Kondratyuk (1897–1942). Both men advanced interesting ideas almost simultaneously yet quite independently of one another.

Fridrikh Tsander   Tsander, a Lithuanian, began his rocket activities in 1908 at the age of twenty-one, six years before he was graduated as an engineer from the Riga Polytechnic Institute. He studied Tsiolkovsky's work, corresponded with him, and edited a compilation of the older scientist's investigations at his request.

Tsander was inspired and even, in part, followed Tsiolkovsky's ideas. But, motivated by his own original concepts, he designed numerous rocket engines, boosters, and rocket planes. He was completely absorbed and devoted to this new science. In fact, he named his daughter, Astra, and his son, Mercury.

Because his ideas were practical and apolitical, he was accepted and admired in the highest academic and political circles, and, unlike Tsiolkovsky, Tsander lectured enthusiastically throughout the Soviet Union on the coming space age. Many of the present Soviet rocket experts associate their earliest interest in rocketry with his lectures.

Tsander generally applied his scientific talent in three areas: 1) the analysis of problems associated with space flight; 2) the development of rocket-engine theory; and 3) the building and testing of rocket engines.

Perhaps one of the more complex problems of the day encountered by researchers working on rocket space flight was the problem of overcoming the earth's gravitational field. Calculations had shown that in order for a single stage rocket to achieve orbital velocity with known propellants, the weight of its propellants had to comprise almost 98 percent of the rocket's total weight. One way of overcoming this problem was by the use of multistage rockets, an approach discussed by such people as Goddard (1919), Tsiolkovsky (1929), and Oberth (1933).

Tsander believed in the multistaged rocket. At that time, however, he felt that it would be most advantageous to use a rocket in conjunction with an airplane. Basically his proposal consisted of an airplanelike device in which the rocket ship served as the body. For atmospheric flights, either a piston engine or a jet engine that used atmospheric oxygen was to serve as the power plant. At high altitudes, the liquid rocket engines were to be started and the metallic parts of the airplane that became unnecessary functionally were to be melted so that they could be used as additional fuel. For the entry phase, to earth or other planets containing an atmosphere, small auxiliary wings were to be used, making it possible to reenter without using propellants. But Tsander realized that the use of metallic propellants would produce only a slight increase in the engine performance. He therefore looked for other

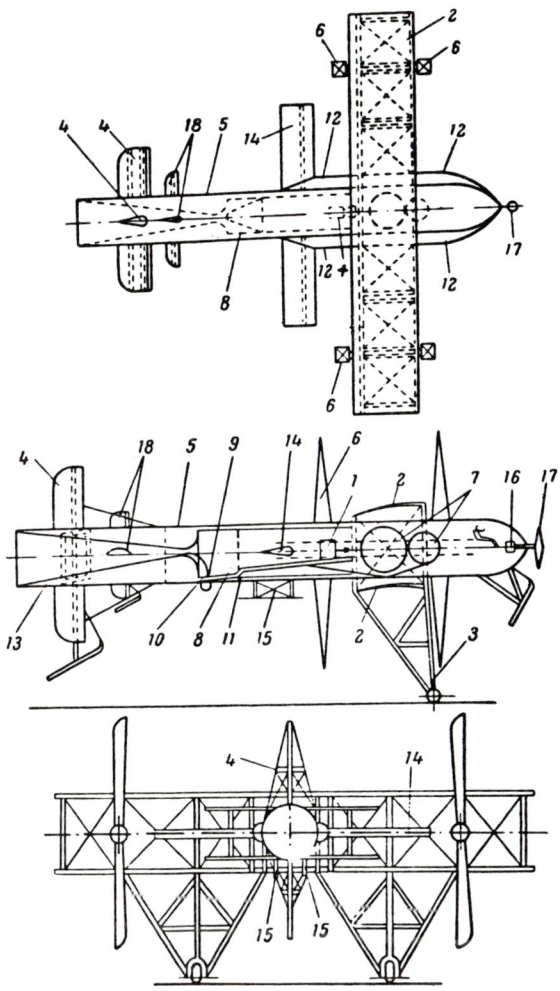

FIG. 16 Interplanetary craft designed by Tsander. Parts of the large airplane: 1) engine, 2) wings, 3) landing gear, 4) rudders, 5) body of the rocket, 6) propellers, 7) tanks, 8) caldron for melting metal parts, 9) molten metal ducts, 10) net to protect the duct from solid metal pieces, 11) fire box, 12) opening for absorbing metallic plane body parts to be ground up and used as rocket fuel additive, 13) nozzle, and parts of the small airplane, 14) wings, 15) landing gear, 16) engine, 17) propeller, 18) rudder.

35

sources of energy that would be applicable to solving the problem of interplanetary flight.

This fact enabled him to differentiate between the types of rocket engines that were most suited to placing a spacecraft into orbit and those most suited to carrying out interplanetary flights.

In the early twenties, Tsander began his theoretical work on the development of rocket engines and by 1928, began to design his first engine, the *Opytnyi Reaktivny* No. 1, or OR-1. The engine was built in the 1929–1930 period when Tsander and his assistants were working at Ts.A.G.I. Later, in 1930, the engine was successfully tested more than fifty times by his group. The engine ran on gasoline and compressed air fed into a combustion chamber from a cylindrical tank. It developed about 11 lb of thrust. The knowledge accumulated from these static firings served as the technical test-bed for designing the OR-2 engine.

Dushkin, Korneev, and Moshkin were among the group of young workers who assisted Tsander in the development of the engine. Testing began on March 18, 1933. The OR-2 engine used benzine and liquid oxygen as propellants and developed a sea-level thrust of about 110 lb. If the tests of the OR-2 proved successful, the plan was to install the engine on an experimental glider known as the R-1.

Ten days after his team first successfully tested the OR-2 engine, Tsander died of typhoid fever at the age of forty-six. His accomplishments and fame were such that he was acclaimed in the Soviet Union as another Tsiolkovsky. In 1959, to commemorate his contributions to rocket development and interplanetary travel, the Soviet Union erected a monument in his honor in the city where he died. It bears the inscription "Fridrikh Arturovich Tsander, Pioneer of Soviet Rocket Engineering," and mounted on the top is a model of the GIRD-X rocket.

Yuri Kondratyuk   The third Russian researcher to be recognized for his advanced concepts in interplanetary flight is Yuri V. Kondratyuk, described by some sources as a mechanic rather

FIG. 17 Yuri V. Kondratyuk, twentieth-century rocket pioneer.

than an engineer. He is perhaps best known for his book, *The Conquest of Interplanetary Space,* published in 1929.

For a period of time, it was believed that this book was his one and only work in this field. However, recently it was discovered that this book was actually his last and most complete work and that at least three manuscript versions on space travel were prepared by him prior to his 1929 book and given to B. N. Vorobyev, a well-known Soviet aviation historian, who had preserved these manuscripts since 1938. Generally speaking, each revised manuscript was an elaboration and improvement on the preceding one. The first manuscript is dated 1916; however, Soviet historians cannot substantiate that date.

Kondratyuk, like Tsiolkovsky, initially worked on a formula to describe the flight of a rocket and came to the independent conclusion, ". . . that the velocity of a rocket's flight in a vacuum depends only on the characteristics of the propellants and on the initial and final mass." After determining that rocket flight to other planets was feasible, Kondratyuk then examined such questions as the effects of gravity and atmospheric drag on the ascent of a rocket, acceleration, trajectories, guidance, stability, and the actual construction of the spaceship. In the same manuscript, he mentions the use of solar energy, employ-

37

ing large mirrors for the purpose, the use of radioactive decay (α and β particles) and cathode rays.

In discussing the different types of energy sources that can be utilized for propulsion systems, Kondratyuk wrote:

> Still another special type of rocket is possible, one that utilizes energy from . . . the sun. In practice, however, this method of operating a rocket is inapplicable at the present time . . . because of purely technical difficulties.

On staging he stated:

> . . . When I speak about the activity of a substance, the weight of this substance should be calculated along with the

FIG. 18   Diagram of Kondratyuk's reentry vehicle with speed reduction by drag: 1) lifting surface, 2) stabilizing surface, 3) pilot's cabin, 4) tail piece, 5) tail.

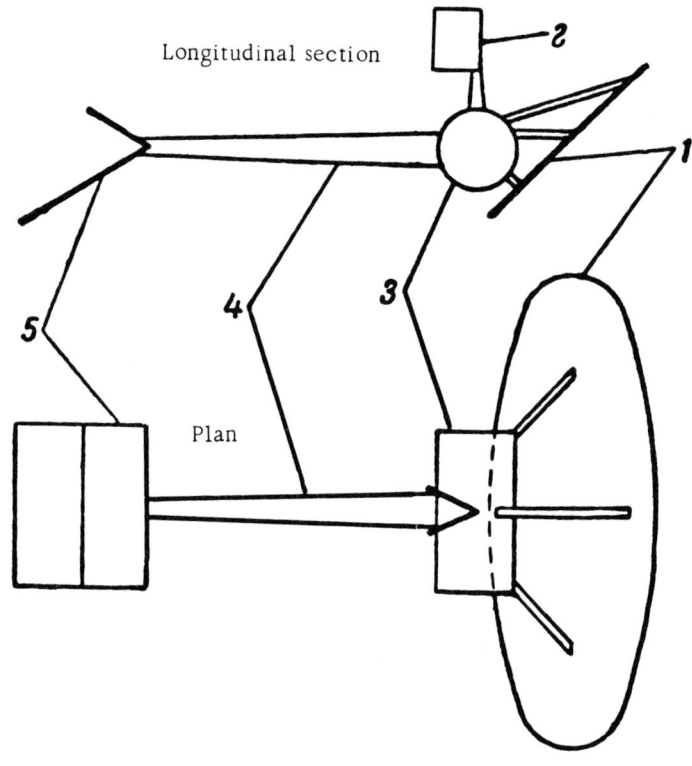

weight of its container; once we have used up a certain portion of the active substance, we reject the tank in which it was carried. It is better, therefore, not to put the entire reserve of active substance in one tank, but to use several progressively smaller ones.

Regarding the methods of reducing the amount of propellant needed for interplanetary exploration, Kondratyuk said:

> In order not to use up a large quantity of active substance, the entire vehicle need not land, its velocity need only be reduced so that it moves uniformly in a circle as near as possible to the body on which the landing is to be made. Then the inactive part separates from it, carrying the amount of active agent necessary for landing the inactive part and for subsequently rejoining the remainder of the vehicle.

This technique is comparable to the Apollo's lunar landing and take-off mode.

In discussing the construction of interplanetary bases, he suggested the creation of such a base in the form of an artificial moon satellite so as to avoid the earth's atmospheric drag effects.

Finally, on returning to earth from orbit or a planetary trip, Kondratyuk suggested two methods of utilizing the atmosphere as a braking medium. The first involved returning the spacecraft close to and then decreasing the speed so that it would circle around the earth, slowly losing speed until it descended to earth as a glider. The second suggested approaching earth without any reduction in speed, using the atmosphere not only for atmospheric braking but also for capturing the vehicle, so that it could not escape from earth again. Kondratyuk admitted that both methods were difficult and dangerous. The latter method was also comparable to the Apollo reentry and return technique.

It should be pointed out that not all the ideas expressed by Kondratyuk can be attributed to him as a "first." The facts are that many of his ideas were originally proposed by other scientists within and outside the Soviet Union. According to Kondratyuk, he did not have an opportunity until 1925 to familiarize himself with the works of other authors, so that he often repeated what had been discovered earlier by others.

Foreign Publications   In addition to the theoretical work on rockets and the development of interplanetary flight, a great deal of effort was expended also on the researching of rocket developments outside the Soviet Union. Starting as far back as 1930, Professor V. P. Vetchinkin was asked to assemble a group of scientists in the Ts.A.G.I. to make a search in foreign literature of the activities being carried out in other countries and to conduct basic research in the field of flight mechanics. The scientists soon translated into Russian and published numerous foreign books on the subject including those by Max Valier and Eugene Sanger.

Soviet Publications   The primary role of disseminating information on astronautics inside the Soviet Union, however, is attributable to a handful of Soviet scientists and nontechnical space writers. Naturally, the foremost among these men was Tsiolkovsky and his many publications spanning almost forty years. His books are not only credited with being informative but also for bringing together all those interested in rocketry and space by including his letters to Soviet aeronautical enthusiasts.

Not as well known, but equally important, was the publication in 1928–1932 of the nine volumes of Prof. N. A. Rynin's *Space Travels*. These books are considered to be an encyclopedia of astronautics, containing the history of the problems of space travel and all the theoretical work on jet propulsion and astronautics known at that time. Rynin, who in Tsarist times was the organizer of the first aerodynamic laboratory in Russia, wrote his monumental work in the coming space age during the most difficult years of Russia's civil war and famine. Recalling his experiences to a friend Rynin said, "I was hungry, I was cold, but one good thing about it—nobody came to see me, nobody interrupted. And so I wrote; it would have been a sin to waste all that time."

About the same time, 1929, Kondratyuk published two books,

entitled *Rockets* and *The Conquest of Space*. Both books were considered to have made valuable contributions to keeping space technology up to date. These were followed in 1932 by Tsander's book, *The Problem of Flight by Jet Propulsion*. Still other valuable literary contributions that informed and stimulated interest in the problems of rocketry and space flight came from popular science books, as well as articles written by such eminent authors as Y. I. Perelman and Prof. K. A. Baev.

The years intervening between 1929 through 1933 were also years of political purges and change in the Soviet Union. Tsander died; Tsiolkovsky was retired and dying from a long illness; and Stechkin was a political prisoner for many years. Eventually he was cleared of all political charges and his position restored. Subsequently, he became a member of the Academy of Science and Head of the Department of Aerospace Propulsion at the Zhukovsky Academy. During the same period, some of the Ts.B.I.R.P. and O.I.M.S. members were also accused of being enemies of the state, and these groups were dissolved.

4 GIRD and G.D.L.

In my experiments I arrive at many new conclusions, but my conclusions are greeted with disbelief by other scientists. These conclusions could be confirmed by repeating my experiments, but no one knows when that will happen. It is hard to work alone for many years under unfavorable conditions and receive neither understanding nor support from anyone.

K. E. Tsiolkovsky, 1931

TSANDER and Kondratyuk were the last of the individual rocket personalities recognized as such within the Soviet Union. Although future space scientists would continue to be singled out for their contributions to the Soviet space effort, the era of group or institutionalized research along functional lines and anonymity would prevail.

As early as 1920, the Soviet rocket pioneers made many interesting and outstanding contributions. However, most of these were concerned with the conceptual or theoretical aspects of rocketry rather than with practical hardware development. This early trend is perhaps best typified by Tsiolkovsky, who in 1928, wrote, "The value of my work lies in the calculations and the conclusions drawn from them. In practical engineering I have achieved almost nothing."

By the end of the 1920s and the early 1930s, however, the difficult years of the reconstruction period were just about over. The nation's industrialization programs were making progress, and the government, recognizing the military potential of the rocket, began to support and encourage coordinated group rocket development. Two of the key organizations receiving this support were the Leningrad Gas Dynamics Laboratory (G.D.L.) and the Group for the Study of Reaction Propulsion (GIRD). Within this compatible environment, the Soviet school of rocketry entered a new phase in its evolution by producing the first of its hardware in the form of experimental engines and flyable rockets.

G.D.L.

The Leningrad Gas Dynamics Laboratory or G.D.L. was formed in 1928 in Leningrad from Vladimir A. Artemev's military powder-rocket group for the expressed purpose of developing high energy, solid rocket motors. Artemev had a long military career that began in the first decade of the 1900s. Both his military and rocket careers got a boost during 1915–1916, when, while still a junior officer, he developed a 3-inch military flare rocket based on an aluminated powder, which revolutionized the state of the art. From that time on his career was devoted exclusively to military rockets.

By 1921, Artemev organized the first Soviet powder-rocket laboratory in Moscow with Nikolai I. Tikhomirov, a rocket designer. Artemev became its first and only director, holding this position until 1927, when the laboratory was moved to Leningrad. In 1928, the laboratory became the G.D.L. of the Revolutionary Military Council (R.V.S.), the name of the Ministry of Defense of the Soviet Union at that time. Tikhomirov, Artemev's coworker, became the fist operational chief under the general supervision of M. N. Tukhachevsky, the head of the Department of Armaments at the R.V.S. In time Tukhachevsky became Marshal of the Soviet Union and Deputy Minister of Defense.

In 1930, Tikhomirov was replaced as the director by B. S. Petrapovlovsky, another military career man. By the end of 1930, Petrapovlovsky's group constructed an 82-mm caliber

FIG. 19 Replica of the ORM-1, the first Soviet liquid-fuel engine.

aircraft rocket and had started research on an antitank rocket weapon.

The ORM* In May, 1929, G.D.L.'s functions were expanded to include the development of liquid-propellant rocket engines. Shortly thereafter, the first series of liquid-bipropellant and monopropellant engines were built in the Soviet Union. On the basis of the theoretical and experimental work conducted by this group during 1929–1930, the first liquid-propellant rocket motor (*Opytnyi Reaktivnyi Motor;* Experimental Reaction Motor), ORM-1, was designed. The engine was built and stand-tested in 1931, producing a thrust of about 44 lb. Its propellants were toluene or gasoline as the fuel and nitrogen tetroxide or liquid oxygen as the oxidizer.

The first monopropellant engine developed by this group was designated the ORM with no numerical model number. A large number of monopropellants were tested in this experimental engine. During 1931 alone, the ORM was fired at least forty-six times. Several innovations that were incorporated into the ORM were an electrode ignition system, a safety valve, and a combustion-chamber pressure switch that automatically shut the engine when the pressure exceeded a predetermined, safe structural level.

The ORM-2 engine was also developed and constructed in 1931. Concurrently, with the engine development Soviet sci-

* ORM and ERM are used interchangeably in Soviet reference sources.

entists emphasized the technology of reliable propellant ignition. Two of the ignition systems tested at that time were a solid-propellant ignitor or squib and liquid hypergolic propellants (propellants that ignite on contact).

As G.D.L.'s capabilities developed in terms of equipment facilities, manpower, and experience, they were able in 1932 to design a block or series of liquid engines. These were designated ORM-4 through ORM-22. These engines were test models used to develop various engine components, such as combustion-chamber propellant injector heads and ignition systems, and to evaluate the correct engine start sequence for various propellant combinations.

In the latter part of 1933, the engine-development test program was accelerated, and G.D.L. within the R.N.I.I. planned a whole new series of liquid-propellant rocket engines running on gasoline and nitric acid. These engines were designated ORM-23 through ORM-102. One engine of the series, ORM-52, developed a thrust of 660 lb and was selected to power a powerful military antiaircraft rocket, a naval torpedo, and a meteorological rocket.

Between 1934 and 1936, the ORM-53 through ORM-66 series of engines were developed. Outstanding in this group was the ORM-65. It was unique in that its thrust level could be modulated or controlled within the range of 110 lb to 385 lb. The ORM-65 like the previous series, utilized kerosene and nitric acid as propellants. Its performance was equivalent to a specific impulse of 210 sec, which was better than the German V-2 and some American missiles that were launched in the late 1940s and early 1950s. Though ORM-65 was unique for its high engine performance efficiency, it was extremely heavy and had a very poor engine thrust to weight ratio, approximately 12. Modern engines have achieved engine thrust to weight ratios of approximately 100.

Rocket development continued at G.D.L. throughout World War II. Between 1932 and 1944, more than 118 liquid-fueled rocket engines of various types were produced by G.D.L., and thousands of static and flight tests were conducted on these engines.

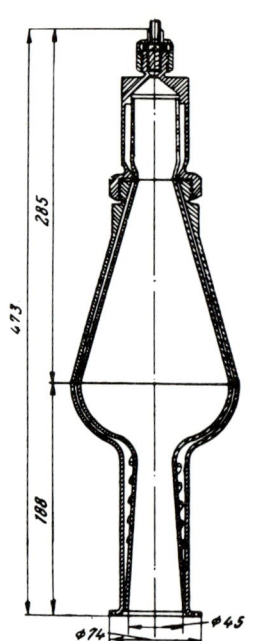

FIG. 20 Alcohol and liquid oxygen engine similar in physical size and performance to GIRD-X engine.

MosGIRD (Moscow Branch of GIRD)

Credit for establishing the foundations of modern Soviet rocketry, in an organizational form, is shared equally between G.D.L. and the various branches of GIRD organizations. G.D.L. is recognized for the development of rocket engines and GIRD for the design, construction, and flight testing of rocket vehicles.

The Moscow branch of GIRD was established in 1931, when a group of Soviet engineers and scientists working on rocket development at Ts.A.G.I. applied to the Chairman of the Central Board of OSOAVIAKHIM (a civil defense organization known as the Society for the Promotion of Defense and Aerochemical Development) for permission to form a special group to conduct rocket studies and experiments within the jurisdiction of that organization. The proposal was approved and a Jet Engine Section was founded, with Tsander as its first director. Within a few months, as a result of the urging of the Society for the Study of Interplanetary Flight the Jet Engine Section was reorganized as the Group for the Study of Jet Propulsion

46

—GIRD. Under the sponsorship of OSOAVIAKHIM, GIRD became national in scale with branches in major cities. The GIRD groups quickly attracted nearly all leading Soviet enthusiasts of interplanetary travel and space research.

GIRD's operation was divided into several technical units: the development of liquid propellant rocket engines; the designing of rockets; the building of a supersonic wind tunnel; and the investigation of problems associated with the application of liquid propellant rocket engines to aircraft.

But just as many of the pioneers who preceded them, this group had their share of ridicule and neglect. For example, in the 1930s food was scarce and rationed in Moscow, and GIRD members were known to have been refused ration books because they were accused of being occupied with "nonsensical fantasies."

In April, 1932, the Soviet government, through GIRD, established the first Rocket Research and Development Center at 19 Sadovo-Spasskaya Street near the center of Moscow. This operation was under the leadership of the young engineer and pilot, Sergei P. Korolyov, the future unidentified chief designer of Soviet boosters and spacecraft. The Korolyov organization

FIG. 21 Chamber of the ORM-65 liquid propellant engine: 1) combustible compound, 2) ignition plug and current collector, 3) electric squib, 4) ignition grain, 5) fuel feed.

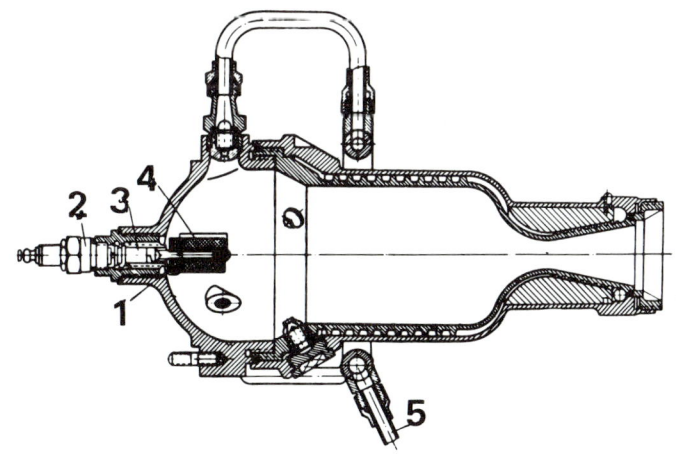

FIG. 22 The 07 rocket with engine located above the center of gravity and propellant tanks located in the stabilizers.

was the prelude to the official history of Sputnik 1 and the manned space flight with Yuri Gagarin.

Initially, the shop at Spasskaya Street consisted of only two semiusable lathes and none of the other tools so essential to rocket work. With the strong support of the government and noted Soviet scientists, however, the GIRD members began their design, development, and construction of rockets. By 1933, the GIRD members had produced four liquid propellant rockets, designated as 05, 07, 09, and 10.

On August 17, 1933, the 09 rocket became the first semi-liquid propellant rocket to be launched in the Soviet Union. At later launchings this rocket obtained an altitude of 4,920 feet. Six such rockets were launched during the period of August, 1933, through May, 1934. The 09 rocket utilized liquid oxygen and so-called "solid" gasoline (a solution of rosin in gasoline constituting a gelatinous mass) for propellants. The propellants were pressure fed to the engine which developed a thrust of about 330 lb and ran for a duration of 15 sec. The rocket was slightly more than 8 ft long, 7 in. in diameter, and had a launch weight of about 40 lb. Of this weight, 11 lb was propellant and about 13 lb was payload.

FIG. 23   Model of the first Soviet rocket (09), launched August 17, 1933. In the background is a model of the final stage of the first cosmic rocket, launched January 2, 1959, which became an artificial satellite of the sun.

Eventually, the 09 rocket became the first antiaircraft rocket in the Soviet arsenal. For the first time in its long evolution, the military rocket made the transition from a solid propulsion to a semiliquid propulsion system. The stage was now set for the development of the longer range, all-liquid-propellant rocket and eventually the intercontinental missile.

The second rocket, the GIRD-X, was launched on November 25, 1933. This rocket is acknowledged officially as the first fully liquid-powered rocket launched in the Soviet Union. The GIRD-X had a streamlined cigar-shaped body with a pointed nose. To maintain aerodynamic stability during flight, it was provided with four oblong stabilizers, almost half as long as the rocket's main body, which was constructed of special structural steel and aluminum alloy. The assembled GIRD-X rocket consisted of five compartments: (1) the uppermost compartment, containing the recovery parachute; (2) next, the oxygen tank; (3) then the compressed air tanks for expelling the propellants; (4) below the pressurant, the fuel tank; (5) and at

FIG. 24   Flight of GIRD-X rocket, the first Soviet liquid rocket, November 25, 1933; the rocket attained an altitude of over 16,000 feet.

the base of the rocket, the propellant feed system and the engine.

LenGIRD (Leningrad Branch of GIRD)

Shortly after the formation of Moscow GIRD in January, 1931, branches similar to the MosGIRD began to appear in other cities throughout the Soviet Union, including Leningrad, Teflin, Kharkov, Arkhangelsk, Novoherkass, and Bryansk. Of these the most active branch was LenGIRD.

Chairman V. I. Shorin and Vice Chairman Chertovskoy, of the Office of Aerial Technology of the Leningrad OSOAVIAKHIM are credited with the initial sponsorship of LenGIRD, and later they gave their support to the group's effort to organize and carry on its rocket work. LenGIRD's first meeting was held in November, 1931, at the Army and Navy House in Leningrad. Officers were elected and the aims of the group discussed. At the second meeting it was decided to form small groups to work and study the various problems of rocket propulsion. Five units were organized: research; planning and designing; public relations; laboratory and launch operations. There were about five or six members in each group. A fund was established to pay for workmen and materials to build and test the rockets and to arrange for technical coordination and liaison between the LenGIRD and MosGIRD operations.

In December, 1931, LenGIRD organized propulsion study seminars and in October, 1932, courses were established for engineering and technical workers where lectures on jet propulsion theory were given by various speakers. To round out the educational program in November, 1932, courses on jet propulsion were offered to persons with elementary or advanced technical education.

In 1932, the planning and design group, headed by V. V. Razumov, designed three solid-motor powered rockets. One of these was a photographic rocket capable of attaining an altitude of about seven miles; another was a flare carrying rocket; and the third an atmospheric sounding rocket.

The photographic rocket's mission was to record the rocket's

FIG. 25 MosGIRD rocket with four parallel propellant tanks in the main body, contrasting to more conventional tandem configuration; built in mid-1930's.

ascent by measuring a base line on the ground that the rocket's camera would photograph from a specified altitude. The camera (weight 11 lb) was placed in the rocket nose and ejected when the rocket reached its peak altitude, automatically taking pictures during the descent by parachute. The body of the rocket was also equipped with a parachute for controlled descent, recovery, and reuse.

Razumov's flare rocket had solid rocket motors that could be mounted individually on the rocket. The number of motors installed at any one given time depended on the desired altitude. The preliminary design of the flare rocket was delivered in early 1932 to Chief of Staff of the Military Artillery Academy. On May 5, 1932, the plans were sent to the research group for the construction of prototypes at the Leningrad Machine Works. In September, 1932, tests were satisfactorily performed at the rocket site on the grounds of the Leningrad OSOAVIA-KHIM. The altitudes reached by the smaller rockets were on the order of three-quarters of a mile.

Of all of LenGIRD's rocket designs perhaps the most unique was the LRD-D-1, a rotary rocket engine conceived by A. N. Shtern in 1932. In the LRD-D-1, the propellants were fed through lines running along the arms on which the rocket engines were mounted. The engine's exhaust nozzles were cut off obliquely so that the jet reactive force had a component in the horizontal plane controlling the lever perpendicularly. Both lever and engines were attached to the bearing of the vertical shaft. This formed a rotary system in which the propellant was fed to the engine under the action of centrifugal forces. The rotating motors not only supplied propellants to the engine but also produced a gyroscopic effect, ensuring stability in flight. The total weight of the rocket was 198 lb. This consisted of a 78-lb structure, 44-lb body assembly, 35-lb engine, and approximately 40 lb of propellants. The thrust of the engine was 440 lb, which gave the rocket an acceleration of 2g at lift-off.

In the article "At the Cradle of Starships," the Soviet writer A. Sachkovo mentions that LenGIRD had no permanent quarters and that work was done in the members' apartments. In special cases when a large conference was scheduled, the meeting was held in a room in the attic of the Leningrad House of Technology.

New people were gradually added to GIRD and their activities were concentrated mostly on increasing the maximum altitude of flights. Calculations were made for rocket ascents up to 30, 60, 100, and even 300 km. In 1932, a plan was worked out for a launch site, the construction of launch pads, observation towers, and underground observation posts with periscopes and photographic equipment.

Unfortunately, political purges of the early 1930s cut the operations of the LenGIRD down to a minimum. By 1935, these operations came to a complete halt. The emphasis was now on military rockets and unclassified literature on rocket research was suddenly withdrawn from the normal channels of distribution. As the Soviets achieved greater successes and began to find more military uses for the rockets, technical information on their construction became even more scarce.

By 1933, the Kremlin, recognizing fully the potential of the rocket and hoping to speed up the development of military rockets, decided to combine the operation of the rocket designers of G.D.L. and the members of GIRD in Moscow. This was done, and a new organization, the Scientific Rocket Research Institute (R.N.I.I.), was created. This organization was staffed by scientists, engineers, and technicians who had now at their disposal the best equipment available. The first head

FIG. 26   Diagram of Razumov's photo-rocket: 1) front compartment with cameras and parachute, 2) gyroscope linked mechanically to rudders, 3) nozzles, 4) stabilizers, 5) rudders.

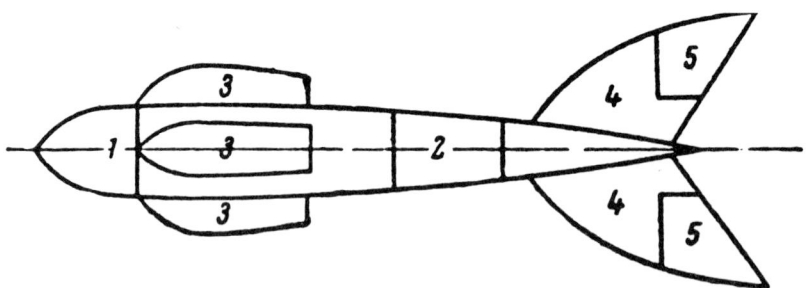

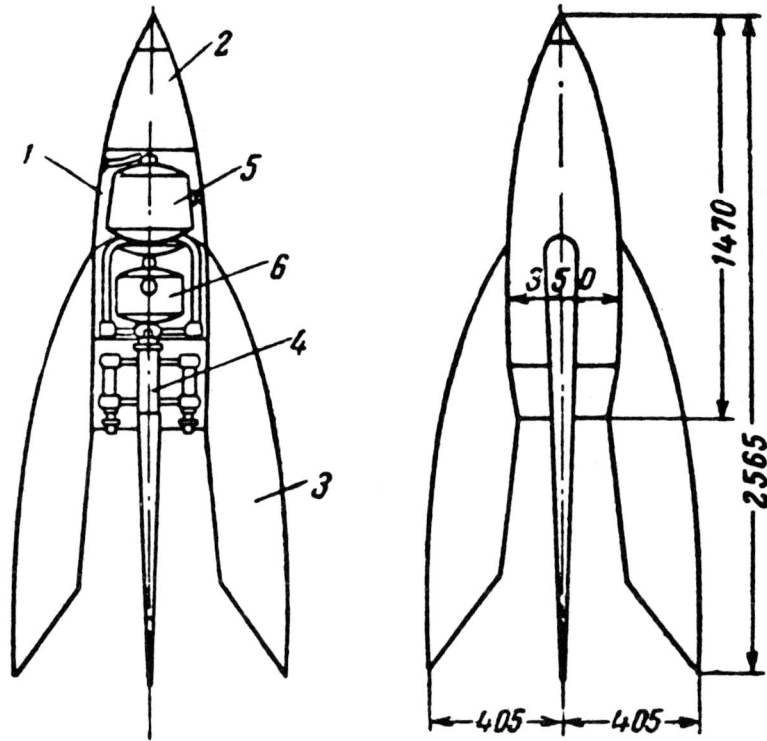

FIG. 27 Razumov-Shtern recording rocket: 1) body, 2) front compartment, 3) stabilizing fins, 4) rotary jet engine, 5) liquid oxygen tanks, 6) benzine tanks.

of the institute was Ivan T. Kleimenov, who wrote to Tsiolkovsky to tell him of the new organization with GIRD membership as its principal researchers. The deputy head was Korolyov.

In 1939, the Soviet government relaxed its security briefly and permitted an article to be published on the successful firing of a two-stage hybrid rocket designed by Igor A. Merkulov. The first stage was a 7.7-lb, solid-propellant motor with a burning duration of 1 sec and the second stage was a 7.8-lb, ram-jet engine with about 5.5-sec supply of propellant. Later in the year another flight took place at night with instruments for precisely measuring the altitude. During this flight, the solid-motor stage first reached a burn-out altitude of 820 ft and a velocity of 447 mph. Between first stage burn-out and second stage ignition there was a lapse of about 2.5 sec in

55

which the rocket rose to an altitude of 2,050 ft. Its speed decreased to 235 mph. At the completion of the second-stage ram jet burn-out, the velocity was increased to 501 mph, and the altitude was 4,329 ft. During the coast phase, the rocket reached a peak altitude of 5,900 ft.

With the start of World War II, all information on rocket development within the Soviet Union was classified, and this policy prevailed until after the launching of the first sputniks.

The years between 1930 and World War II were both extremely critical and the most productive for the infant Soviet space program. These were the years when the concepts proposed by men like Tsiolkovsky and Tsander became a reality, the years when initiative and originality were put to the test. These also were formative years, when the ideals and higher aspirations of men and their dreams of flights to the stars were gradually subordinated to the military objectives and requirements of the nation.

5 The Men of GIRD

> . . . I understand better than any-
> one else the abyss that separates an
> idea from its realization, since in
> my course of life, I have not only
> thought and calculated, but have
> also completed projects working
> with my own hands. . . .
>
> *K. E. Tsiolkovsky*

OUT OF the GIRD and G.D.L. groups rose three men who have been identified as playing major roles in the development of Soviet spacepower. They were Sergei P. Korolyov, Valentin P. Glushko, and Mikhail K. Tikhonravov.

Sergei Korolyov

Sergei Korolyov was born on December 30, 1907, in the Ukraine and died suddenly at the age of sixty on January 14, 1966. His obituary confirmed what was long suspected, that Korolyov was the "Chief Designer of Carrier Rockets and Spacecraft," a title used by the Soviets for almost ten years to shield his identity. Korolyov's obituary said in part:

> In 1933, he participated in the organization of the GIRD, which created the first experimental rockets. He was the outstanding designer of space systems, by which, for the first time in the world, artificial earth satellites were launched, a Soviet emblem was landed on the moon, and a flight was made around the moon to photograph its reverse side. He directed the creation of piloted spaceships, in which man for the first time in history made a flight into outer space and took a walk in space.

Korolyov began working in the aircraft industry as a test pilot and a young designer of early Soviet planes and gliders.

57

In the spring of 1931, Korolyov took part in setting up the jet Engine Section of the OSOAVIAKHIM. Tsander was selected as the first director of this pioneer group which shortly thereafter was reorganized into GIRD.

In April, 1932, Korolyov took the lead of GIRD's design and production section and launched a large-scale rocket-development program. He dedicated himself to the building of rocket-propelled space systems. All his work begun at GIRD was continued at R.N.I.I.

His first book, *Rocket Flight in the Stratosphere*, was published by the Ministry of Defense in 1934. In the years that followed, he worked in the Ministry of Aircraft Production, in the Ministry of Armaments, and in the Rocket Research Institute, always in charge of rocket research and development.

Another technical area that interested Korolyov was the design of aircraft and the application of rocket engines to aircraft both as primary and auxiliary propulsion systems. Two of the projects that have been associated with him are the design of the 212 experimental winged rocket and the glider SK-9.

The 212 aircraft was powered by the ORM-65 rocket engine, which was developed in Glushko's operation. Its thrust was variable, between 110 lb and 385 lb, and its propellants were kerosene and nitric acid. Between 1937 and 1939, many successful flights were flown in this aircraft. These are acknowledged as the first rocket-powered aircraft flights conducted in the Soviet Union. Some Soviet sources say that Korolyov participated in some of these flights as the test pilot.

Korolyov's other aircraft-design project, the SK-9 glider, was built in 1939 under his direction. It was powered by an engine developed by L. S. Dushkin, which unlike the ORM-65, had a fixed thrust of 150 lb. Little else is known about the program except that in February, 1940, a series of successful test flights were completed.

Korolyov's activities in developing a rocket assist take-off (RATO) for aircraft evidently were confined exclusively to the World War II period. The records, brief in content, state that "during World War II Korolyov designed liquid-fuel rocket accelerators for fighter planes and dive bombers and himself took part in the test flights." After the war, Korolyov resumed practical work on rocketry and the application of rockets to outer space research, occasionally combining his work with lecturing at colleges. In 1945, he was given the responsibility for developing advanced versions of the German V-2 rocket.

By 1947, the Soviets produced thousands of V-2s, developing advanced techniques for mass production and for training their rocket corps. By 1949, the T-1 or Pobeda, the Soviet version of the V-2, was in production for strategic and scientific applications. And it was in 1949 also that authorization for a 2,000-mile IRBM was approved. This missile has been referred to as a modification of the German A-4/A-9 and at various times has been identified as the T-2 and the Model 103.

The T-2 was successfully tested and put into production by 1952; it set the stage for the development of the T-3 or the ICBM. Acceptance and authorization of the ICBM (Vostok) design came in 1954; it was a design submitted by Korolyov's group.

Just a few years later his team developed a rocket that was first a military-weapon delivery system; then the carrier rocket for launching the world's first artificial earth satellite on October 4, 1957.

Korolyov's activities did not stop with the development of the carrier rocket. He also was instrumental in the construction of recoverable vehicles for man's flight into space and for piloted spacecraft of the Vostok and Voskhod type, which safely carried the world's first manned flight into space and were also used for the first walk in space.

Under the guidance of Korolyov, the Soviet Union built and launched its first interplanetary automatic stations to Venus in February, 1961, to Mars in November, 1962, as well as interplanetary automatic stations of the Zond series in June, 1965. The Zond 3 was launched into a Mars trajectory in the first few days of its flight. Zond 3 photographed sections of the back of the moon and transmitted the pictures to earth. Korolyov also participated in the development of the first Soviet communication satellite, Molniya. He shared in designing automatic lunar stations, which ensured soft-landing of research instruments on the moon. These instruments transmitted panoramic pictures of the moon's surface to earth for the first time.

For his many printed works, inventions, and designs Korolyov was elected Corresponding Member of the U.S.S.R. Academy of Artillery Sciences in 1947. In 1953, the same title was conferred on him by the U.S.S.R. Academy of Sciences, which elected Korolyov a full member Academician in 1958. For a number of years, Korolyov was a member of its Presidium. In addition, he earned the title of Hero of Socialist Labor twice, won the Lenin Prize, and received many other Soviet decorations.

V. P. Glushko

The 1959 Yearbook of the Great Soviet Encyclopedia contains the following entry for Glushko:

> Glushko, Valentin Petrovich, born 20 August, 1906, is a Soviet scientist and a specialist in the field of heat engineering; he has been an Academician since 1958, a Member of the Communist Party of the Soviet Union since 1956, and a Corresponding Member of the U.S.S.R. Academy of Sciences since 1953. His work pertains to various problems of heat engineering.

Glushko's name is synonymous with the early history of liquid rocket-engine development in the Soviet Union. This fact is confirmed by the well-known Soviet writer, N. Rynin, who suggests that Glushko began his professional career in 1924 at the age of sixteen with the publication of "The Conquest of the Moon by the Earth," followed in 1926 by "Station beyond the Earth."

Glushko is recognized by many within Russia as the man who transformed Tsiolkovsky's rocket-engine theories into practical engine hardware. Specifically, he is singled out as the designer of the first liquid-propellant rocket engine, the ORM-1. Chronologically, the essential research and design studies were conducted between 1929 and 1930. Most of the engine's ninety-three parts were developed in 1930, and the engine was assembled and tested in 1931. For its static test firing the engine was installed in an upright position, that is with the rocket's exhaust nozzle pointed upwards. In this position, the combustion chamber with its adjacent tubing was enclosed in a water-filled sleeve for cooling. Evidently the engine had the capability of burning at least two different propellant combinations, toluene or gasoline for fuel and nitrogen tetroxide or liquid oxygen as the oxidizer. The engine produced a thrust of 44 pounds. The Soviets claim that the firing of the ORM-1 was the first liquid-propellant rocket engine "in the full sense of the words" fired in the Soviet Union.

Test data on engines ORM-4 through ORM-22 gave Glushko essential information to confirm his theoretical analysis of propellant and engine performance, as well as evolving the most suitable engine arrangements for liquid-propellant rockets. Furthermore, the test data led to the formation, in 1932, of the basic principles for reliable rocket engine ignition and shockless combustion methods for liquid rocket systems. In 1933, Glushko's research team successfully developed the so-called regeneratively cooled combustion chamber, which then made long-burning-duration engines feasible. Prior to this development, liquid engines were either coated with heat-resistant materials or were equipped with water-cooling jackets, both unsatisfactory. With the introduction of regenerative cooling techniques, former test problems were essentially eliminated. Regeneratively cooled engines used one of the propellants (like liquid oxygen) to cool the engine's combustion-chamber wall and nozzle-throat area, usually in a double-jacketed wall arrangement.

With the solution of this problem, Glushko and his group next turned their attention to designing more powerful and sophisticated engines, including those to be used for aircraft. Before the end of 1933, Glushko developed his second series of engines (ORM-23 through ORM-52) with which he continued his engine experimental programs, accumulating more than one hundred test firings.

Between 1934 and 1936, Glushko's group developed the third

series of engines (ORM-53 through ORM-66). One of these engines the ORM-65 was throttleable and scheduled for installation in a flying bomb and a rocket airplane designed by Korolyov.*

Between 1941 and 1946, Glushko's group switched to the development of a series of thrust-augmentation rocket engines for aircraft, designated the RD-1, RD-2, and RD-3—developing thrusts of 660 lb, 1,320 lb, and 1,980 lb respectively.

After about four hundred test firings of the RD-1, it was installed on such combat aircraft as the PE-2, the LA-7R, and the Yak-3. The RD-1 was superseded by the RD-2 and, by 1945, by the RD-3, which was equipped with a gas generator and turbine pump.

Glushko's activities after World War II are not fully known, but numerous analysts have associated him with the engine development for the first ICBM, the Vostok, and later generations of carrier rockets. Because of the unusual sensitivity of his position, the Soviets have tried for a number of years to disassociate him from liquid-propellant rocket-engine developments, both past and present. In the Soviet press, he is identified only as the "Chief Designer of Rocket Engines."

According to a United States government report published in 1966,

> . . . Glushko followed Korolyov into oblivion in 1959, a year after they both had been elected full members of the Academy of Sciences. His name is conspicuous by its absence in recent histories, although it is listed on the masthead of the house organ of the Academy of Sciences, the Vistnik. Glushko is now laboratory chief at the Moscow Institute of Mineral Fuels; he also holds posts in Voronezh, 300 miles south of Moscow, at the state university and the Voronezh Logging Institute.

Mikhail Tikhonravov

The third functionary of the Soviet space program, identified by title again and not by name, is the "Chief Theoretician of

---

* Soviet references did not specifically identify Glushko as the man responsible for the ORM engine developments, but the material presented here represents my conclusions based on research and reconstruction of events.

Cosmonautics." As with Korolyov and Glushko, this man has been identified by many sources outside of Russia as Mikhail K. Tikhonravov, but his identity has never been revealed by Soviets.

Tikhonravov was born in 1900. In the foreword to the second edition of his book, *The Flight of Birds and Machines with Flapping Wings*, Tikhonravov says of himself: "I began work on the investigation of birds in flight in 1922. My article 'Some Statistical and Aerodynamical Data on Birds' was published in 1926, and, by 1939, I had published a total of eight articles on the subject of flight aerodynamics in the periodical *Aircraft*."

Rynin states that in 1924 Tikhonravov wrote an article, "Density of Air and Its Change with Altitude," for a military journal. In addition, he includes Tikhonravov in the list of Russian scientists studying interplanetary communications. In "Super-Aviation and Super-Artillery," Rynin presents several formulas for determining the relative density of air, and among these he lists one by Tikhonravov. This reference seems to indicate that, at least by the early 1920s, Tikhonravov was actively engaged in work on aerodynamics and in research on interplanetary communications.

Tikhonravov, like Korolyov and Glushko, was a member of the original group around Tsiolkovsky in the later twenties and early thirties. As a member of GIRD, Tikhonravov took a leading part in its operations. He is credited by many Soviet authors as the designer of several of the early rockets developed in the Soviet Union. Foremost of these, he is recognized as the designer of the semi-liquid-propellant rocket that was successfully launched in Russia in 1933, the 09 that reached an altitude of 1,300 feet.

At the same time, it appears, Tikhonravov was also working on the development of a liquid-propellant meteorological and antiaircraft rocket. In August, 1933, successful tests were conducted on the ORM-50 engine developed by Glushko and his group. Between 1933 and 1934, the ORM-50 was installed in Tikhonravov's rocket and in 1934–1935 the rocket was successfully launched several times. After each launching, Tikhonravov continued to improve his design until in 1935 his rocket reached an altitude of more than 32,800 feet, or six times as high as his 09 rocket. The ORM-52 engine may have been used in the advanced rocket versions.

From its performance, it is obvious that this rocket was significantly superior to anything built at that time. Its design was classified. Yet in 1945, Soviet scientists, going through the ruins of Germany, found drawings of this rocket in the archives of the Nazi Air Ministry. There is no explanation as to how these plans got into German hands.

Tikhonravov's activities during World War II are not discussed in available Soviet literature. Like Korolyov and Glushko, he is directly associated with the development of their space program. For example, Yuri Gagarin and Gherman S. Titov describe the Theoretician of Cosmonautics as in charge of all programs to compute the flight parameters of the Vostok series, including automatic computer control.

In addition to his work with rocket development and flight performance, Tikhonravov continued to contribute to the technical literature of his day. In 1935, he wrote a book on rockets as well as several technical articles, and after the war he edited books written by Tsiolkovsky and Tsander. In 1951, Tikhonravov forecast in "Flight to the Moon" that two men could fly around the moon and back to earth on a rocketship weighing approximately 1,000 tons.

From 1954 on, Tikhonravov began to be set apart from the usual group of design engineers and soon became more closely identified with great scientists. In 1957, the Presidium of the U.S.S.R. Academy of Sciences established the Commission on Interplanetary Communications (I.C.I.C.) to "coordinate and direct all work concerned with solving the problem of mastering cosmic space." Academician L. I. Sedov was appointed chairman, and Tikhonravov was appointed vice chairman.

After 1959, nothing signed by Tikhonravov has appeared in the press. His name like Korolyov's and Glushko's has disappeared from the lists of important members of research institutes in the Soviet Union. However, the first issue of the publication *Space Research* in the summer of 1963 carried his name on its masthead.

The total membership of GIRD, including all branches throughout the Soviet Union, and G.D.L. exceeded 1,000 members. Some of the top people in these groups were the young space enthusiasts of the 1920s and 1930s who had dedicated their lives to the development of rocketry and space travel. They had now become the backbone of the Soviet rocket technical movement. The GIRD and G.D.L. roster of the early 1930s includes many other names that have since become an integral part of the Soviet space program. Some of these men are discussed throughout various sections of this book; many others are not. For the record, a more complete list of the rocket specialists of GIRD and G.D.L. follows.

MEMBER	TECHNICAL SPECIALTY
Artemev, V. A.	Solid rockets
Dushkin, L. S.	Rocket aircraft and engines
Glushko, V. P.	Liquid rocket engines
Gryaznov, V.	Engine development
Isoyev, A. V.	Engines
Korneev, L. K.	Combustion experiments
Korolyov, S. P.	Aircraft, boosters and spacecraft
Kostikov, A. G.	Solid fueled military rockets
Langemak, G. E.	Solid rockets
Merkulov, I. A.	Ram jet
Moshkin, E. K.	Engine development
Perelman, Y. I.	Space author
Petrapovlovsky, B. S.	Solid propellant rockets
Pobedonostev. Y. A.	Rocket booster
Polyarny, A. I.	Combustion experiments
Razumov, V. V.	Booster design
Rynin, N. A.	Space author
Salikov, A. V.	Engine development
Shtern, A. N.	Engine design
Tikhonravov, M. K.	Rocket engines and vehicles
Tsander, F. A.	Rocket pioneer
Tsiolkovsky, K. E.	Rocket pioneer

6 Prelude to Sputnik

> The twin shells of gravity and the atmosphere restrict the rocket to the vicinity of the earth's surface. Pierce them and you will be a denizen of outer space, able to direct your rocket toward any planet or asteroid.
>
> *K. E. Tsiolkovsky, 1931*

THE SOVIET policy of keeping its rocket scientists anonymous and its rocket developments under strict military security began in the early 1930s. But starting with the latter part of the 1940s and early 1950s, the news blackout policy became more relaxed in certain areas. Since then, a limited amount of information has become available through Soviet publications, which allow us a glimpse of some of the events that transpired during the earlier years.

It is now well known that Russia's military commanders have always been interested in the rocket for military purposes. Lesser known, however, is the Soviet scientists' early appreciation of the rocket's potential as a research tool. Since the 1920s, they have been interested in investigating the density and temperature of the upper atmosphere, the intensity of solar and cosmic radiation, and the concentration of charged particles. Furthermore, a number of medical, biological, and other high-altitude experiments interested them that could only be performed by rocket flights.

Research in the Later 1930s

In 1935, two conferences on upper atmospheric research were held. The first took place in Leningrad and the second in Moscow. At the Leningrad meeting, which was convened by the Academy of Sciences of the U.S.S.R., it was decided that, in view of the great future of jet propulsion, it was necessary to concentrate on mastering the techniques of manned rocket flight. At the Moscow conference, resolutions were made to initiate the construction of an experimental winged rocket for manned flight at low altitudes and a stratospheric rocket for scientific research. The papers presented at these conferences were later published in a book entitled *Reaction Motion,* which, at that time, could be obtained by anyone in the Soviet Union or abroad.

World War II Rockets

During World War II, G.D.L. and GIRD concentrated primarily on developing the highly effective multibarrel Katusha, or Little Katy, ground-to-ground rocket. Katusha batteries are credited with turning the tide of the war by stopping the Nazi armies at Moscow, Leningrad, and other fronts. The man given credit for the development of the Katusha missile is A. G. Kostikov. Actual work in his project started in about 1936, and by 1940 the Katusha missile passed all qualification tests and was accepted for mass production. As its inventor, Kostikov was awarded a Stalin Prize of 100,000 rubles.

No other rocket project came to fruition during the war due to the fact that the main Soviet industrial centers were occupied by the German armies or evacuated to the east, and all available resources were being directed to the immediate defense of the country.

Shortly before the end of World War II, the United States, its Allies, and Russia realized that Germany was far more advanced in the development of large, guided missiles. Accordingly, they organized teams of technical experts, both civilian

and military, to search the German research centers and fac-
tories for information pertaining to rocket development. Liter-
ally tons of captured documents and all types of rocket hardware were taken as war booty.

German research   Real interest in rockets was evidenced in
Germany first in 1929, with the formation of the Society for
Space Travel (*Verein für Raumschiffahrt*). An abandoned 300-
acre arsenal on the outskirts of Berlin was used as a proving
ground for some simple missiles they had built. In the spring
of 1932, a black sedan drove up to the edge of the proving
grounds, and three passengers, one of them General Dornberger, got out to watch a rocket launching. "They were in
mufti, but mufti or not, it was the army," von Braun said
later. "That was the beginning. The Versailles Treaty hadn't
placed any restrictions on rockets, and the Army was desperate
to get back on its feet. We didn't care much about that, one
way or the other, but we needed money, and the army seemed
willing to give it to us."

The army supported the members of Society for Space
Travel and provided them generously with equipment, proving
grounds, and skilled workmen, and by 1936 the coastal site of
Peenemunde had been selected as an experimental station. Hitler, on coming into power, allocated 20 million marks to Peenemunde and granted high priority to materials and labor assigned to the project. By 1937, the station was completed with
Major General Dornberger designated the military head and
von Braun the civilian director. The operation was charged
with developing a rocket field weapon capable of carrying
large warheads beyond the ranges attainable by artillery.

With Hitler's successes in Poland and France, however, the
Luftwaffe became his pet organization, and he decided that
he no longer needed guided missiles to win the war. As a result, Peenemunde's priorities practically disappeared, technicians were refused military deferment, and work on the A-4
(V-2) was drastically curtailed. Seeking support for their
program, Dornberger and von Braun visited Hitler in the

summer of 1942. They were told that Hitler had dreamed that rockets would not work.

By autumn the Luftwaffe had failed to subdue England and the invasion of Russia was bogged down at Moscow and Stalingrad. When a V-2 on October 3, 1942, was successfully launched 120 miles down range, Hitler revoked his dream. The second visit of Dornberger and von Braun brought a promise of his support and a demand that the V-2s be readied for mass production.

In January, 1943, a committee recommended the establishment of three assembly plants, the Southern, the Central, and the Eastern. The Southern Plant was to be midway between Vilnva-Neustadt and Friedrichshafen. The Central Plant was to be situated in the southern Harz Mountains in the vicinity of Nordhausen; and the Eastern Plant was to be located in Latvia near the city of Riga.

In the spring of 1943, as mass production plans of the V-2 were nearing completion, two missile battalions were organized, one mechanized and one in a bunker installation. The mechanized battalion was to have a launching capability of twenty-seven V-2 missiles per day. It was divided into three batteries, with three launching platforms and nine missile trailers each. The bunker battalion was to be established in Cherbourg. This facility was to have a launching capability of thirty-six V-2s per day; a storage capacity of 108 missiles and three-days' fuel supply.

In May, 1943, the Peenemunde activities attracted the attention of the British Intelligence Service. After observing their operations for some time, the British, on the night of August 17, 1943, launched 600 planes to bomb Peenemunde. This raid resulted in 815 German casualties, almost complete destruction of the test stands, hangars, and loss of transportation networks. Even so, mass production of V-2s began only one month after the September deadline set by Hitler. By late September, the production of V-2s for research purposes reached about twenty missiles per month.

In June, 1944, the launch facility in Cherbourg was captured by American forces; by July, 1944, the Eastern Plant had fallen into Soviet hands; and the plants in Vienna and Friedrichshafen had been damaged by Allied air raids to such an extent that only partial assembly could be performed. Thus, the Central Plant was left to carry out the major part of the assembly work.

In addition to the raids on the prime assembly plants, Allied raids severely crippled subcontractor facilities, particularly in the Stettin and Stuttgart areas, and additional bottlenecks in ball-bearing deliveries were caused by bombing Schweinfurt.

On September 6, 1944, the first tactical V-2 was launched against England by a mobile unit from a site near The Hague. In the period from August, 1944, through February, 1945, 3,000 A-4s were delivered to the field units. Inspection of the first 1,000 delivered revealed that 339 were defective. About 5 percent of the remaining 661 failed to leave the launch stand. Others exploded on the launcher and some tumbled after take-off. By October, 85 percent of the missiles received by the V-2s crews were successfully launched. The majority of these reached the general area of the target, but only 20 percent reached the specific target. By the end of the war, a total of 3,300 V-2s had been launched.

In January of 1945, Soviet military forces reached Pomerania and were threatening Peenemunde. Preparations were made to destroy everything that could not be removed to a new location. A total of almost five thousand people plus equipment was evacuated to the Central Plant. A crash program was conducted to increase the accuracy and eliminate the air bursts of the V-2. Plans also called for the production of 600 missiles by September, 1945.

These plans were never completed. The Soviets captured Peenemunde. The United States Army occupied the Harz Mountains, capturing the Central Plant.

Just prior to this, about five hundred of the key guided-missile personnel and technicians were taken south of Munich by SS troops to be executed rather than to allow their being captured. The movements of the Allied forces were so rapid that the Nazis did not get time to carry out their plan. On March 15, 1945, orders were received from SS headquarters to remove and hide all documents to prevent their capture. Two members of the Peenemunde group carried out these orders by loading all available material on three large trucks and trailers and taking it to an abandoned mine in the Harz Mountains. The documents were unloaded in the mine, and the entrance was blasted shut. Their efforts were in vain. By the end of April, 1945, United States Army units located the mine and found the hidden documents.

After V-E day, Magnus von Braun, the brother of Wernher von Braun, contacted the United States forces and told them that a very large number of the Peenemunde scientists were living in small villages throughout the Bavarian Alps and that many of them wanted to surrender to the United States. Approximately one hundred fifty of the engineers, scientists, and technicians were rounded up and offered five-year contracts to bring their missile know-how to the United States.

The United States' roundup of German rocket scientists, secret technical documents, and missile hardware was conducted under the code name Operation Paperclip, and a comparable project by the British was designated Operation Backtrack. The Soviets' code name for their operation was never made public.

Soviet acquisitions   On May 6, 1944, the Second White Russian Army under Gen. Konstantin Rokossovsky, occupied Peenemunde. To salvage what they could of the bomb-damaged installation and round up the dispersed rocket specialists, the Russians implemented their equivalent of America's Operation Paperclip. Their search was not in vain. They found the infamous V-2 (*Vergeltungswaffe-Zwei;* Vengeance Weapon Number Two), blueprints of the ocean-spanning A-9/A-10 rocket, and the surface-to-surface Rheinbote; also the air-to-air R-4/M rocket capable of carrying a pound of explosives, the air-to-surface HS-293, FX-1400, and the V-1 buzz bomb, as well as the experimental missiles called R-10, R-12, and others. In the surface-to-air category, the Soviets got the Rheintochter, Wasserfall, Taifun, and the series known as the Henschel missiles in various stages of development.

The Soviets in addition to Peenemunde found other rocket centers either partially or completely destroyed. Most importantly they learned that German rocket scientists General Dornberger, Wernher von Braun, and some one hundred fifty key members of their team were offered jobs by and evacuated to the United States.

The Soviets on the other hand got Helmut Groettrup, an electronics expert. Groettrup in turn gathered some two hundred men to go with him in key positions to Peenemunde to rebuild its facilities. Within a short period of time, he rebuilt the V-2 production capability, as well as a series of laboratories, shops, and test facilities known as the *Zentralweike,* or Central Works. Within a year's time Groettrup's operation totaled 5,500 men.

Stalin, however, was not satisfied with the progress of recruiting and rebuilding of the German rocket complexes and rockets and ordered Gen. Ivan Serov of the secret police to move the entire operation into the Soviet Union. In October, 1946, more than six thousand German technical specialists of all kinds and about twenty thousand members of their families were gathered without notice, put on waiting trains with their belongings, and shipped to various research centers throughout the Soviet Union.

In a later assessment of the German rocket capability, the Soviet scientists felt that in the field of originality and theory, the Soviet Union was not behind Germany, and, in some respects, they were even ahead of the Peenemunde group. Because of this, the Soviets were less apprehensive about the top echelon of the German rocket scientists taking their skills to the United States. In the field of practical rocket engineering of the V-2 size, however, the Soviets conceded that they were not as advanced as the Germans.

Post-World War II Rockets

Rather than the idea men, what the Soviets needed most was the practical, technological experience and advanced missile-production know-how, which they hoped to acquire from Groettrup's group.

Dr. Tokaty, writing about the events leading up to the development of a Russian intercontinental missile, stated:

> Marshall Zhigarev, the Commander in Chief of the Soviet Air Force, said to me in 1946, "We must admit that our V-2–type rockets do not satisfy our long-term needs. Should there be an American–Soviet war, they would be useless; what we really need are long-range rockets capable of hitting target areas on the American continent. This fact should dominate the mind and efforts of your rocket group." In 1947, at a meeting in the Kremlin of aircraft and rocket designers, G. M. Malenkov said, "I am not happy with our V-2s; we cannot rely on such a primitive weapon; our strategic needs are predetermined by the fact that our potential enemy is to be found thousands of miles away."

One day later, at a meeting of the Politbureau and of the Council of Ministers of the Soviet Union, I. V. Stalin emphasized the goals:

> Under Hitler, German scientists have developed many interesting ideas. This "Sanger Project" seems to represent one of them. Such a rocket could change the fate of the war. It could be an effective strait jacket for that noisy shopkeeper, Harry Truman. We must go ahead with it, comrades. The problem of the creation of transatlantic rockets is of extreme importance to us.

Accordingly, Stalin personally suggested, and the Council of Ministers agreed immediately, to the formation of a special State Commission for the study of the problems of long-range rockets (*Pravitel'stvennaya Komissiya po Raketam Dalnego Deistviya;* P.K.R.D.D.). It consisted of Col.-Gen. I. A. Serov (First Deputy Minister of N.K.V.D.; Chairman), Prof.-Col. G. A. Tokaty-Tokaev (Chief Scientist and Deputy Chairman; from the Soviet Air Forces), Prof. M. V. Keldysh (Ministry of Armaments), Prof. M. A. Kishkin (Ministry of Aircraft Production), and Maj.-Gen. I. V. Stalin (member). This decree signified a turning point in Soviet rocket research. Toward the end of 1947, everybody wanted to design a transatlantic rocket.

FIG. 28 Russian-built V-2
being tested in the Soviet
Union.

At that time, the state of rocket technology in the Soviet Union was roughly as follows:

1. The problem of small military rockets had been solved.
2. Rockets of the V-1 and V-2 types were already in production.
3. Rockets of the "Pobeda" type were in design.
4. There were two draft projects for long-range rockets. One of them, known as "project TT-1," was presented to the government in September, 1947. It was a three-stage, liquid-fuel rocket for extremely high altitude and orbital flights.

In the immediate postwar period, Kostikov, the developer of the Katusha, was assigned to the Kubyshev rocket center and German scientists under his direction carried out major rocket experiments. Yuri A. Pobedonostev, a former member of the Ts.A.G.I. staff and RNII-No. 1 since the late thirties, was also given charge of a large group of German engineers to build and eventually improve on the performance of the V-2.

By 1945–1946, V-2 production was completely restored in the Soviet Union, and by 1946–1947 it surpassed the German's production level of 1944. The exploration of the upper layers of the atmosphere by the V-2 rockets began in the autumn of 1947. By 1948, Pobedonostev improved the German V-2 and went into production in 1949 with the Soviet version known as the Pobeda. The Pobeda had a range of 560 miles, more ample than that of the V-2. From 1949 on, upper atmospheric research was continued with the Pobedas. Then in 1950–1951, the first Soviet Rocket Division was formed and armed with V-2s and Pobedas.

Research in the Fifties

By 1949, the Soviets produced single-stage geophysical rockets that could attain an altitude of 68 miles with an instrumented payload of 286 lb. This permitted the Soviets to get valuable data on the structure of the atmosphere up to these altitudes. They recovered by parachute containers of test instruments and, later, experimental animals.

At a meteorological conference that was held in Washington in October, 1957, a Soviet representative delivered a technical paper in which he described the two-stage meteorological rocket that had been in production within the Soviet Union since 1950. He stated:

> It consists of a solid-propellant booster rocket about 4.5 ft long and weighing 517 lb that burns 180 lb of powder in 2 sec and a 23-ft-long sustainer rocket, having a starting weight of 1,496 lb and a kerosene and nitric acid engine that develops a thrust of 3,014 lb for 60 sec. At an altitude of 43 miles, the sustainer rocket separates into two parts, the upper part with instruments attaining an altitude of 56 miles. Both parts descend by parachute and are recovered.

American sources have since identified this rocket as the MR-1 or Mitio and have listed its performance as a 176-lb

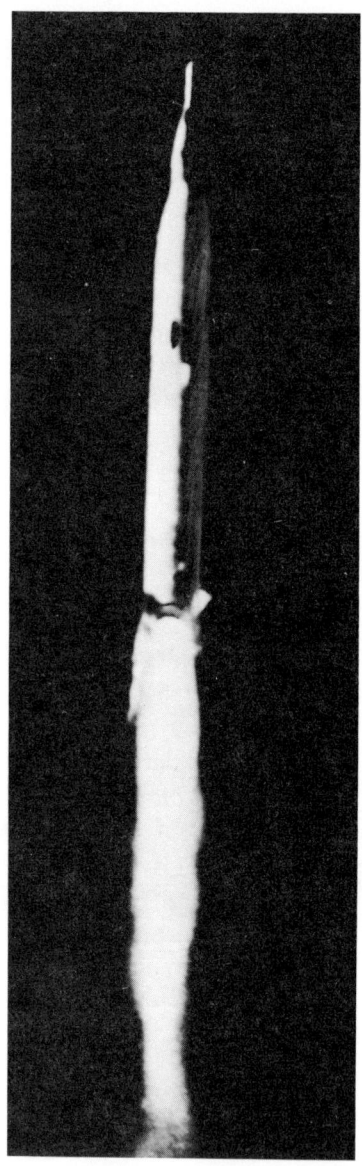

FIG. 29 *Left:* Early 1950 Soviet geophysical rocket; instrument containers can be seen projecting at the sides. *Right:* Geophysical rocket launched February 21, 1958; instrument container carried in nose.

payload carried to a 120-mile altitude. It was acknowledged as the work-horse rocket during the IGY.

According to Academician A. A. Blagonravov, 1951 was also the year when the successful biological research flights using dogs convinced scientists of the feasibility of manned rocket flights. Also, at this time, Soviet scientists were convinced they could place heavy payloads into orbit. Accordingly, preparations for these flights were started with the development of space suits, ecology, and emergency escape systems.

With the successes of their ballistic missile programs, exceptional progress in their meteorological and biological experiments, and a more relaxed security policy, there were many more statements being made by highly placed and highly respected members of the scientific community regarding their ability to build a rocket to launch an earth satellite.

In recognition of the work being accomplished by their rocket people and to show the endorsement of the Soviet hierarchy of their activities, the Presidium of the U.S.S.R. Academy of Sciences established the K. E. Tsiolkovsky Gold Medal in 1954 for outstanding work in the field of interplanetary communications, to be awarded every three years beginning in 1957. In the same year, a major milestone in developing a satellite launch capability was reached when the second design for an ICBM, which was submitted by Korolyov's group, was accepted for development. Two years later, several Soviet ICBMs were ready for testing.

Progress in their atmospheric research program was parallel and comparable to the progress in their ICBM program. The records indicate that by May, 1957, a single-stage rocket ascended to an altitude of 132 miles with a payload of 4,840 lb. In reality these rockets were the real precursors to unmanned and manned orbital flight, the altitudes achieved by the meteorological rockets were just at the perigee altitudes of the first three sputniks (140 miles) and the payloads carried by the meteorological rockets (4,840 lb) was of the magnitude that permitted the Soviets to check out just about all of the instrumentation that was needed for these satellites prior to the orbital flights.

August 27, 1957, saw the culmination of the work started by such men as Tsiolkovsky, Tsander, Korolyov, and Glushko, as well as by organizations such as GIRD, G.D.L. and RNII when a Tass report in *Pravda* stated that "successful tests of an intercontinental ballistic rocket and also explosions of nuclear and thermonuclear weapons have been carried out in conformity with the plan of scientific research work in the U.S.S.R."

The impact of the ICBM launching and its connection with a satellite was still not obvious to most people outside the Soviet Union. At the completion of the August tests, the Russian people learned that their nation had completed a series of ICBM tests that sent military payloads higher than 800 miles to targets more than 5,000 miles away.

In September, the Soviet people were again informed by radio that their country was getting ready to launch two different types of satellites. There was considerable speculation that this historical event would occur on September 17, 1957, the one hundredth anniversary of the birth of Tsiolkovsky. Then, just a few days before the launch, Radio Moscow confirmed the Sputnik 1 radio-transmission frequencies. These announcements essentially went unnoticed outside the country.

FIG. 30   Geophysical instruments after landing from altitude of 294,-000 feet.

7 Sputnik

Mankind's first great step forward
into outer space consists in flying
beyond the atmosphere and creating a satellite of the earth. The
rest is comparatively easy, even
escape from our solar system. But of
course, I do not have in mind, here,
descent on to the giant planets.

K. E. Tsiolkovsky, 1926

THE CLIMAX to one of man's greatest technological chronicles occurred on October 4, 1957, when Sputnik 1 was established in orbit, and *Pravda* announced the successful launching of the world's first artificial satellite.

During the next two and one-half years, eighteen months of
the International Geophysical Year (I.G.Y.) and twelve months
of the International Geophysical Cooperation (I.G.C.), the Soviets continued to demonstrate the prodigality of their rocket
program by launching, in addition to their sputniks, 175 meteorological rockets. According to plan, 125 of these were
launched during the I.G.Y. and 50 more were launched in
1959 during the I.G.C. program.

The sophistication of their geophysical program, achieved
by 1959, is perhaps best indicated by their announcement reporting the firing of a single-stage research rocket to an altitude of 280 miles, carrying a payload of 3,726 lb. The payload
included geophysical scientific equipment, radio telemetry
devices, power sources, and a sealed cabin with auxiliary systems containing two dogs. The dogs were recovered by means
of a parachute system.

Soviet scientists were apprehensive that the initial earth
satellites might end in failure because of an incorrect judg-

79

ment of the physical quantities that were to be measured. Success was highly dependent on the measuring range of the satellites' instruments, which in turn were sensitive to the on-board environmental temperature.

It was not known what temperature could be maintained in an orbiting satellite. It might overheat; or the temperature might fall below the permissible level, causing the instruments to record incorrectly or break down completely. Furthermore, equally intricate and directly conflicting requirements often arose. These included, for example, the selection and priority of experiments, the specific orientation of the on-board equipment, and the characteristics of Sputnik's orbit. Resolution of most of the problems required many compromises with the final selection predicated on obtaining the maximum experimental results from Sputnik's flights.

In designing the satellite and its associated equipment, the following ground rules were established and followed:

1. Sputnik had to be an all-purpose probe, not just designed for a specific application.

2. Equipment had to be designed for multiple applications.

3. Since other satellites to be launched after Sputnik 1 would use the same system, the telemetry had to have a large capacity to accommodate increased information requirements.

4. Requirements for other aspects of future spacecraft development had to be taken into consideration where possible.

Fortunately, for the Soviets, payload weight was not a problem, as sufficient booster power existed.

Sputnik

The satellite configuration that evolved was a sphere with a diameter of 23 in. and weight of 184 lb. Its shell was a sealed casing made of an aluminum alloy, with the outer surface polished and suitably treated for internal temperature control. All equipment, including the power supply, was located within the satellite. Before launching, the satellite was filled with gaseous nitrogen.

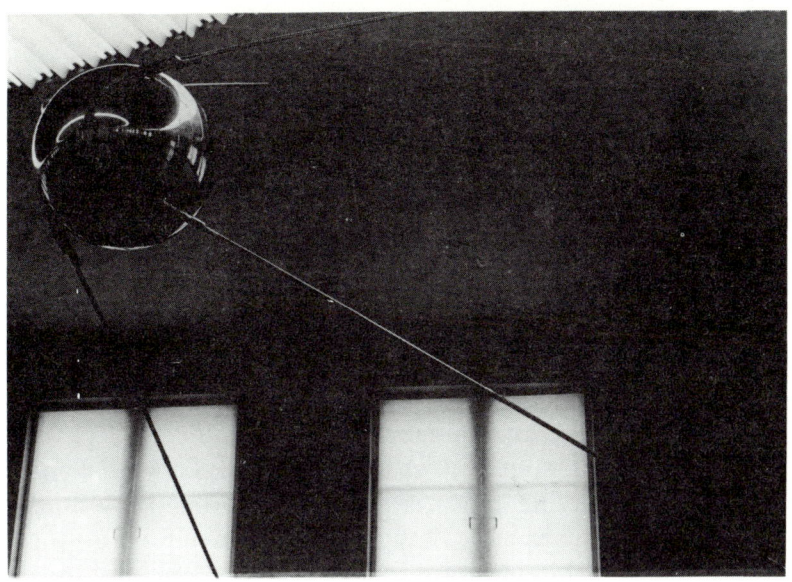

FIG. 31   Sputnik 1, launched October 4, 1957; weight was approximately 185 lb, diameter 23 in.

During the launch and orbital injection phases, Sputnik 1 was housed in a protective nose cone in the forward section of the upper portion of the injection stage. After the upper stage reached an altitude of 142 miles and a velocity of nearly 26,240 ft per sec and was flying almost parallel to the earth's surface, the engine was shut off. The protective cone was jettisoned and the satellite, separated from the orbital injection rocket, began its free flight in orbit around the earth. The satellite reached an apogee of 588 miles and had an orbital period around the earth of 96 min. Its orbital inclination to the equator was 65°. The moment the satellite separated from the injection rocket, pivot antennae 8 to 9.5 ft long opened to their working positions. Two radio transmitters, operating on frequencies of 20.005 and 40.002 *mc* provided the communication link to earth.

The Sputnik's carrier rocket remained in orbit for sixty days and the satellite for ninety-four days, allowing ample time to conduct experiments that were heretofore impossible. For example, Sputnik 1 obtained the first information about the physical conditions in the upper atmosphere and the temperature inside a spacecraft in orbital flight. The temperature was monitored by means of temperature sensors that produced changes in the frequency of the telemetry signals. This information was essential so that future flights could incorporate proper temperature control systems for the on-board instruments. The information was also required as a check of the heat-regulation system and for the proper preparation of future biological experiments that are nearly always temperature sensitive.

To attain the much needed upper-atmospheric data, Sputnik 1 was utilized in two different ways. First, the aerodynamic characteristics of spherical shapes are well known. Consequently by observing the orbital changes, primarily altitude, resulting from atmospheric drag, it was a relatively simple matter to calculate the density of the upper layers of the atmosphere in the region of the orbit perigee. Second, since the carrier rocket was also in orbit with the satellite, a similar experiment was conducted by observing the orbital-decay rate of the rocket itself.

An equally important experiment conducted by Sputnik 1 was the investigation of the ionosphere by observation of the propagation of the 20.005 and 40.002 mg radio waves emitted by the satellite. Radio signals of these frequencies are propagated through the ionosphere in a curve rather than a straight line. Measurements of the difference in time between the satellite's optical and radio rising and setting made it possible to determine the curvilinear distortion of the radio beam and, thereby, the electron content of the atmosphere in the path of this beam. The nature of the propagation of radio waves in the ionosphere was also studied, since this was very important for ensuring reliable communication with space vehicles.

Subsequent Launchings

Three months later, on January 31, 1958, America launched her first satellite, Explorer 1. It transmitted ionospheric data for nearly four months and is credited with discovering the

Van Allen radiation belt. In contrast to the 184 lb of Sputnik 1, Explorer 1 weighed only 31 lb.

In rapid succession, a family of increasingly larger unmanned orbital, planetary, and manned test spacecraft were launched. The first ten Soviet space launches were all referred to as sputniks before various program names developed. Actually *sputnik* is the Russian word for "traveler," and the name has since come to refer to any Soviet satellite.

Sputnik 2   The Sputnik 2 was launched on November 3, 1957, carrying research instrumentation and a living passenger, the dog, Laika. It weighed 1,118 lb, almost a thousand pounds more than Sputnik 1. It was approximately 12 ft long and 6 ft in diameter. Its orbital characteristics were apogee 1,038 miles, perigee 140 miles, period 104 min, and an orbital inclination of 65°. This launch was again a surprise because of the quick follow-up to the first launch, the extremely heavy payload weight, and the fact that it carried a living animal.

Launching dogs in rockets was not a new experience in the Soviet Union. Before 1957, medicobiological experiments carried out with high altitude and geophysical rockets had provided quite extensive information about the ability of dogs to endure the accelerations and stresses, as well as the state of weightlessness during the coast phase of the flight. However, the duration of weightlessness in those flights was much too short to be used in calculations having to do with orbital flights.

Sputnik 1 proved it possible to provide a suitable on-board temperature for Laika. A new system had to be devised, however, to furnish Laika with breathing oxygen. In a weightless condition, the natural circulation of gas is impossible. Therefore, a system of a forced ventilation was installed in Laika's airtight capsule. Highly reactive chemicals provided the oxygen the dog required for breathing and absorbed the carbon dioxide and water vapor released inside the capsule. Laika remained alive for over a week within this controlled environment, flying approximately one hundred orbits around the earth, and revealing for the first time the long-term effects of weightlessness on a living animal.

Information on Laika's flight was continually monitored and her condition relayed to earth via telemetry. From the results of this flight, Soviet scientists came to the conclusion that animals could endure lengthy weightlessness and other conditions of long orbital flights. In particular, the results of the investigation of the influence of solar and cosmic radiation on a living organism, made with Sputnik 2, provided information pertinent to protecting future astronauts from the dangerous effects of radiation not encountered in the terrestrial atmosphere.

The physical experiments carried out with Sputnik 2 were in some measure similar to those made with Sputnik 1—the determination of the density of the atmosphere; the study of ionosphere; and the monitoring of temperature changes inside the instrument container, which was the exact replica of the first satellite in design.

One of the most important results obtained from the physi-

FIG. 32   Laika in pressurized cabin before being placed in satellite.

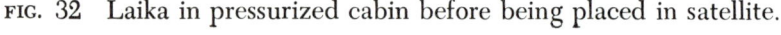

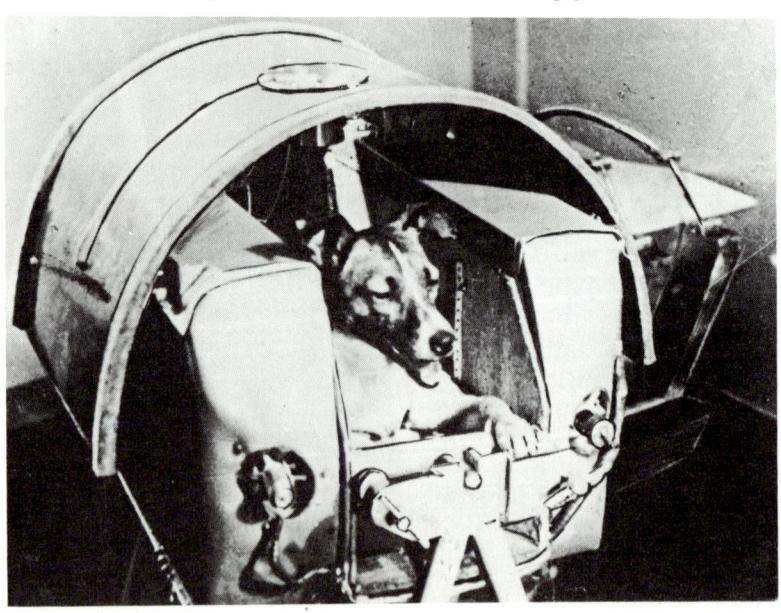

FIG. 33 Sputnik 3.

cal experiments aboard Sputnik 2 was the discovery that as
the orbital altitude and geomagnetic latitude (over which the
satellite was passing at any given time) increased, there was a
subsequent increase in the flux intensity of the high-energy
charge particles. It was later discovered that this was because
the satellite entered the Polar regions of the outer zone of radiation now known as the Van Allen belt.

Sputnik 3 On May 15, 1958, Sputnik 3 was launched into orbit. The satellite weighed 3,018 lb, about three times that of
Sputnik 2. Of this payload weight, 2,129.6 lb were allocated
for experiments, instrumentation, telemetry, heat-control sys-

85

tem, and power supply. The primary types of experiments consisted of measuring the earth's radiation belts, the sun's corpuscular radiation, upper atmospheric density, properties of the ionosphere, the earth's magnetic fields, cosmic rays, and the impact of micrometeorites. To obtain this information, the satellite was placed in an orbit with a 141-mile perigee, a 1,168-mile apogee, an orbital period of 106 min, and an orbital inclination of 65°.

The survey of high-energy particles showed that when the satellite entered the radiation belt between 55° and 65° latitude, both north and south, there was a sharp increase in the drag exerted on the satellite. This occurred because of the bombardment of the satellite by electrons having energies of 100 kiloelectron-volt (keV) or more. It was also established that the radiation intensity increased with altitude. As with Sputnik 2, these phenomena were the result of the satellite's passage into a zone now known as the outer Van Allen radiation belt.

With the instruments designed to measure the sun's corpuscular radiation, electrons with an energy of 10 keV were directly detected for the first time. Fields of these electrons were measured between the altitudes of 290 miles to 1,168 miles. The electron intensity was greater by day than by night and changed constantly with greater altitude or higher geomagnetic latitude.

By knowing the satellite's orientation, the direction of movement of these particles in these fields was determined. It was discovered that as a rule the particles moved in a direction perpendicular to the earth's magnetic lines of force. The speed of these electrons was observed to be much faster than that of the solar corpuscles, determined by observation of the auroras. It was suggested that these were evidently atmospheric electrons accelerated in the outer atmosphere by the transitory geomagnetic fields.

Measurements to determine the density of the upper atmosphere were made by observing the changes in the satellite's orbital parameters. The values of the density obtained by observations of the first three satellites proved to be higher by a factor of 10 than those calculated. These measurements were taken in a period of maximum solar activity, however, when the upper atmosphere is highly charged ("hottest" from a radiation standpoint) and extends farther into outer space. Subsequent investigations in a less active solar period confirmed the "cooling" of the atmosphere resulting in a lowering of the electron density in the upper atmosphere.

To measure the earth's magnetic field at the orbital altitude, a self-orientating magnetometer (an instrument for measuring the intensity of the earth's magnetic field) was included in the instrumentation. The magnetometer was also used to determine the orientation of the satellite at any given moment. This was an important piece of information, necessary for interpreting other measurements. Other on-board instruments measured the intensity of primary cosmic rays, determined the quantity of heavy nuclei in cosmic radiation, and detected micrometeorite impacts.

Manned Spacecraft

Between May 15, 1960, and March 25, 1961, the Soviet Union launched five spacecrafts, Sputniks 4, 5, 6, 9 and 10, designed for manned use.* The primary objectives of the manned precursor launches was to flight-test all of the spacecraft systems, study the flight trajectory, and conduct additional medico-biological investigations in preparation for manned flight.

Sputnik 4 The first of five "manned" prototype test spacecraft was launched on May 15, 1960. It weighed 9,988 lb, had a perigee of 198 miles, an apogee of 229 miles, a period of 92 min, and an orbital inclination of 65°. The primary objective of this flight was to test the spacecraft's orientation and deorbit systems. According to official U.S. sources, recovery of this spacecraft failed when the deorbit system improperly boosted the spacecraft into a higher orbit. It carried a robot only and burned on reentry.

* Sputniks 7 and 8 were unmanned planetary probes.

Sputnik 5   On August 19, 1960, the second "manned" prototype test spacecraft, Sputnik 5, was launched with the dogs,
Belka and Strelka, on board to conduct additional biological
tests. Its weight was 10,120 lb, and its orbital parameters
were apogee 210 miles, perigee 190 miles, period 91 min and
orbital inclination 64°. Sputnik 5 was the first spacecraft to
be recovered. It landed on August 20, 1960, some twenty-four
hours after launch.

Recovery of the spacecraft made it possible to carry out,
along with other physical experiments, a study of cosmic rays
by means of nuclear photoemulsions. In this experiment, cosmic
ray particles passing through the nuclear photoemulsions col-
lide with atomic nuclei, causing the nuclei to disintegrate and
produce new particles. The newly formed particles undergo
a series of changes. The secondary particles themselves react
with the atomic nuclei of the emulsion. By studying the photoemulsions under the microscope, it was possible to reconstruct
the picture of processes taking place in thousand-millionths of
a second. Sputnik 5 carried several blocks of thick-layer nuclear
photoemulsions, one of which was developed in flight. This was

FIG. 34   Strelka with second generation.

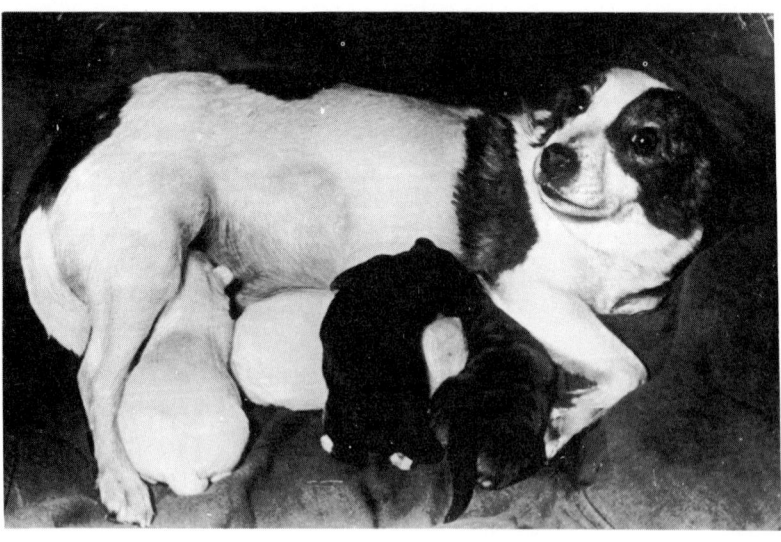

a more reliable means of insulating the tracks of individual nuclei against the general background of cosmic radiation.

Sputnik 6 Sputnik 6, the third in the series of "manned" spacecraft tests, was launched on December 1, 1960, and landed after twenty-four hours of orbital flight.

The spacecraft weighed 10,038 lb. Its perigee was 116 miles, its apogee 163.3 miles, its period 89 min and its orbital inclination 65°. The flight test program included systems development, space research, medical and biological experiments on the dogs Pchelka and Mushka, and on other small animals, insects, and plants. According to official U.S. sources, recovery failed, with the dog passengers perishing on reentry.

Sputnik 9 Sputnik 9, the fourth "manned" test spacecraft was launched on March 9, 1961. It weighed 10,340 pounds, had a perigee of 114 miles, an apogee of 155 miles, a period of 89 min, and an orbital inclination of 64° 65 min. The payload included the dog Chernushka and a dummy in the cosmonaut's seat. The test included, as in the other launches, the development of on-board systems, scientific space research, medical and biological experiments. The spacecraft reentered after one orbit.

Sputnik 10 Sputnik 10 was the fifth and last launch in this series, before the manned launch occurred on March 25, 1961. Like the preceding test, the spacecraft reentered after a single orbit. The weight, orbital parameters, and tests were also comparable. This launch carried the dog, Zverdochka, and a dummy in the cosmonaut's chair.

Interplanetary Probes

Sputnik 7 On February 4, 1961, the Soviet Union implemented a new phase in its space program by launching the first of its interplanetary probes, Sputnik 7, into orbit. This launching was the initial attempt at sending probes to Venus. The first of these two Venus sputniks failed to be launched from its parking orbit. The satellite burned on reentry on February 26, 1961, after about three hundred seventy revolutions around the earth.

Sputnik 8   Eight days later, on February 12, 1961, Sputnik 8 was successfully launched from a parking orbit. Subsequently, radio contact with Venus 1 was lost after the spacecraft had traveled 4.7 million miles.

These were the first steps in unraveling the mysteries of the universe. The research accomplished with the first three satellites came at the peak of an eleven-year cycle of solar activity. In order to study further the dependence of the parameters of the upper atmosphere and the interplanetary medium on solar activity (and to accumulate data during the period of minimum solar activity for further statistical processing), it was necessary to have systematic flights, systematic space investigations, and efficient observation of the phenomena taking place. This task was assigned to the spacecraft that would follow the sputniks, in ever increasing numbers, throughout the remainder of the decade.

8 Genealogy

> What we usually observe of jet propulsion on earth is so pitiable as to encourage nobody to dreams and research. Only intellect and science could indicate the transformation of these phenomena into what is grandiose and almost inscrutable to the senses.
>
> *K. E. Tsiolkovsky, 1911*

WITHIN the decade after the Soviet Union first launched Sputnik 1 into orbit, more than two hundred thirty of its automatic scouts have orbited the earth, moon, and sun. Still others have landed on extraterrestrial soil, the moon and Venus. According to the official Soviet count, during the first ten years of the space age, they had orbited earth, lunar, and solar satellites weighing in toto 1,500 tons (3 million lb), including the weight of the last stages of the carrier rockets that entered the same orbits—an average weight of 14,000 lb per launch. And, as planned, cosmonauts have followed their automatic scouts into orbit, and one can surmise that they are preparing manned flights to follow their scouts to the moon and later to the planets nearest earth.

Launch Vehicles

Although the space goals of the United States and the Soviet Union appear to be comparable, their schedule and technical implementation are not the same. This is particularly true in the case of large boosters. To a large extent, the early successes of the Soviet space program is attributed to the exploitation of

FIG. 35   Vostok booster which launched the first man into orbit April 12, 1960.

their military boosters, particularly the application of ICBM and MRBM classes of launch vehicles.

The ICBM-Vostok was used in four basic versions. Originally, it was used in its unmodified ICBM form for the first sputnik launches. Then as the payloads got heavier and the missions more complex, the ICBM versions were modified by improving the performance of the existing hardware and adding new and additional stages, in tandem, to the basic ICBM. For smaller payloads of the Cosmos and other series of spacecraft, two other military boosters, the Skean and Sandal, were probably modified. In payload-carrying capability, the Skean is comparable to our Atlas and the Sandal is comparable to our Thor. Completing their stable of launch vehicles are two rockets, the Proton, which is in the same payload class as our Saturn 1B, and the Lunar rocket, which is reported capable of placing larger payloads into earth orbit and on the moon than Saturn 5.

The exact identity of the early space booster remained a mystery for almost ten years. Then, at the Paris air show in 1967, the Vostok was displayed to the public for the first time. It was unlike any of the drawings and descriptions projected by analysts. The basic configuration was revealed as consisting of a central core stage surrounded by four large tapered strap-on stages. Furthermore, it was shown that the core and each strap-on unit had four engines, with each pair of engines fed by a single turbopump, and that thrust-vector control (booster steering) was obtained by twelve smaller control nozzles using the exhaust from the pump turbine.

Designed before the thermonuclear breakthrough, the ICBM-derived launch vehicle is estimated to have a first stage approximately 67.5 ft long with a maximum diameter of 33.8 ft at the base. The propellants are liquid oxygen as the oxidizer and kerosene as the fuel. The sea-level performance of each of the twenty main engines is a 55,000-lb thrust with an estimated specific impulse of 250 sec. The total lift-off thrust (including the attitude control nozzles) is about 1 million lb, and the orbital payload of the basic ICBM version is on the order of 3,000 lb. Because of the confidence, reliability, payload capability, and other inherent advantages normally obtained with repeated use of any space system, the Soviet Union did not change the basic booster (central core and strap-ons) but concentrated on the upper stages for fulfilling more sophisticated missions.

Although the Soviet's first ICBM launch vehicle proved cumbersome as a military system and overpowering as a nuclear delivery system, it proved to be ahead of its time as a

CHART II.   Soviet Launch Vehicle Genealogy

Genotype	Derivatives

ICBM

| ICBM Launch | Sputnik 1, 2, 3 | | | 2 stages Combined F 1,130,000 lb 5 engines 3,000 lb wt in No. 3 |
| 8/57 | 10/57–5/58 | | | |

| Lunik 1, 2, 3 | | | | 3 stages Combined F 1.3 lb 6 engines 3,300 lb engine |
| 1/59–10/59 | | | | |

| Sputnik (Korabl) 4, 5, 6, 9, 10 | Vostok 1–6 | Large cosmos reconnaissance | Probably Polyot series | 3 stages Combined F 1.3 lb 6 engines 6 engines New upper stage |
| 5/60–3/61 | 4/61–6/63 | 4/62–Present | 11/63–4/64 | |

| Planetary Luniks | Voskhod Multiman & EVA | Cosmos life support | Soyuz family | 4 stages Combined F 1.4 lb 7 engines |
| 1/63–12/66 | 10/64–3/65 | 2/66 | 4/67 | |

Sandal 1959

MRBM & IRBM

| Small cosmos Misc. scientific, & military satellites |
| 3/62 |

Skean 1961

Small (1,200 lb or less) and medium sized (7,500 lb or less) cosmos satellites are probably launched on Sandal MRBMs and Skean IRBMs, respectively. A new common upper stage using the RD-119 engine is probably added to the basic vehicle.

| Medium cosmos NAVSAT |
| 8/64–5/67 |

Proton & Lunar

? — Configuration unknown. Sources indicate the vehicle was designed by a new group (not Designer Korolov's) and is totally unlike its predecessor, Vostok. All engines are still developed by V. P. Glushko's Gas Dynamics Laboratory.

? — Unclassified sources suggest the possible existence of a new Soviet launch vehicle larger than the Saturn V.

space booster in terms of payload capability and building-block concept for improving its performance as a space booster. This was demonstrated in 1959, when the Soviet space plans called for the first lunar launches. Although the basic ICBM sputnik did not have the characteristic velocity to meet this new requirement, a second-generation launcher was developed simply by adding a stage, in tandem, on top of the ICBM-sputnik launch vehicle.

In December, 1959, this three-stage booster was used to launch three probes to the moon. These probes each weighed over 795 lb and were designated Luna 1, 2, and 3. Thus the ICBM-sputnik launch vehicle utility was extended from the very inefficient first applications simply by the addition of an upper stage. This extension improved its gross payload-carrying ability to about 12,000 lb in earth orbit or over 3,000 lb to the moon.

The versatility of the Soviet space booster was again demonstrated in their unmanned precursor and manned launch program. This time "product improvement" of the third-stage engine resulted in an increase in the specific impulse from an estimated 300 sec to 314 sec. In terms of payload, its earth-orbit capability was increased to over 13,600 lb, lunar to about 3,500 lb, and more than 1,600 lb to the planets.

The first application of the booster was the launching of the Korabl sputniks (five manned precursors in the sputnik series). The six manned flights that followed were also boosted by the Vostok. It was from this series of manned flights that the name Vostok was generally applied to the family of boosters developed from the first Soviet ICBM.

While the Korabl sputniks and Vostok flights were achieving their share of "firsts" in space exploration, Soviet scientists were planning to launch even larger payloads into earth orbit and to the planets. Once again the building-block concept was exploited. Now the three-stage version of the Vostok vehicle was modified by the addition of another stage. As a fourth stage vehicle, the Vostok had an earth-orbit payload capability of over 17,000 lb, which made it especially useful for planetary flights. The new member of the Vostok family was first used in 1963 to launch a series of Mars and Venus probes. The next year three cosmonauts, Komarov, Yegorov, and Feoktistov, rode the new launch vehicle into earth orbit, capturing another "first" for the Soviet Union—the first multimanned flight.

Table 1
Soviet Launch Vehicle Specifications

Name	Description	Thrust (lb × 10⁶)				Orbital payload (lb)	Lunar payload (lb)	Planetary payload (lb)
		1	2	3	4			
ICBM Sputnik	2 stages (1 1/2 as in Atlas)	1.13				3,000		
Lunik (Luna)-Vostok	3 stages (added upper stage)	1.13	0.20			12,000	3,000	
Vostok, manned	3 stages (improved upper stage)	1.13	0.20			14,000	3,500	1,600
Vostok-Voskhod/Soyuz	4 stages (added 4th stage)	1.13	0.20	0.1		18,000	3,600	2,500
Sandal L/V	Sandal MRBM with added upper stage	Thor/Agena equivalent				1,200		
Skean L/V	Skean IRBM with added upper stage	Atlas/Agena equivalent				7,500		
Proton	New vehicle unlike Vostok	3	1	Configuration unknown		27,000 (60,000)		
Lunar L/V	Advanced state of development	10.0M total				325,000	125,000	Mars 50,000 Venus 60,000

Notes 1 Lunar and planetary spacecraft are launched from a 14,200-lb orbital launching platform.
      2 Performance data are best available estimates.

As Soviet technology advanced in orbital spacecraft, a series of smaller 1,000-lb to 5,000-lb scientific satellites were developed and flown. The Vostok standard launch vehicle was too large for these payloads, and consequently smaller boosters had to be developed for these launches. The Soviets had earlier produced a substantial arsenal of MRBMs and IRBMs. To meet the new space requirement they followed the precedent set by Vostok and modified two other military missiles for space applications. The missiles selected are believed to have been the Sandal and the Skean. The Sandal (Thor class) is a single-stage MRBM with an estimated characteristic velocity capability of about 12,500 ft per sec. The Skean is a two-stage IRBM (Atlas class) with a characteristic velocity of about 17,000 ft per sec. Clearly, neither was adequate for orbital applications from an energy standpoint. To fulfill the orbital payload requirements, a new upper stage, probably using the RD-119 engine, was developed and added to each booster. This made it possible to place about 1,200 lb to 7,500 lb into low earth orbit with the Sandal and Skean boosters, respectively.

The earth's largest artificial satellite was launched into low earth orbit in July, 1965. This was Proton 1, referred to by the Soviet news media as an orbiting physics laboratory. Two more followed it within a year. In referring to the Proton's launch vehicle, several Soviet leaders stated that it was a brand new, powerful vehicle. Most observers agree that although the Proton vehicle launched only 27,000 lb, it had a much greater capability. Its capability, after a "product improvement" cycle and the addition of upper stages, will be near 60,000 lb in low earth orbit. If uprated, the Proton could launch a manned capsule on a circumlunar flight, but could not carry sufficient payload for a manned landing mission.

An actual manned landing on the moon requires a much larger booster than the Proton. The U.S. booster Saturn 5 is designed to have a low earth-orbit payload of 250,000 lb. Any Soviet booster must necessarily be this large or larger. Unclassified sources suggest the possible existence of a new Soviet vehicle larger than the Soviet Proton and the U.S. Saturn 5. It should be noted that, as of the latter part of 1969, the Soviet lunar launch vehicle has not been launched; nor, for that matter, has its existence been acknowledged by the Soviet Union. In view of the Soviets' boast that they would welcome U.S. astronauts on the moon, however, it is reasonable to assume that they are belatedly developing such a capability.

Four families of Soviet space boosters have been developed,

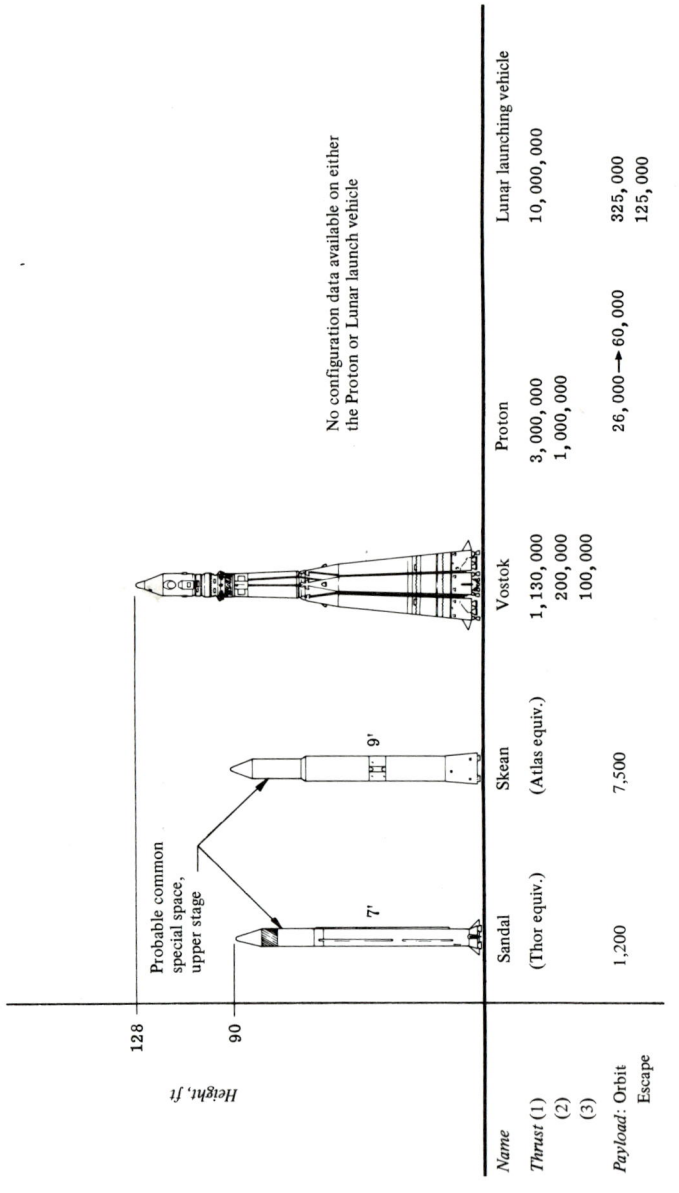

Name	Sandal	Skean	Vostok	Proton	Lunar launching vehicle
Thrust (1)	(Thor equiv.)	(Atlas equiv.)	1,130,000	3,000,000	10,000,000
(2)			200,000	1,000,000	
(3)			100,000		
Payload: Orbit	1,200	7,500		26,000 → 60,000	325,000
Escape					125,000

No configuration data available on either
the Proton or Lunar launch vehicle

Probable common
special space,
upper stage

7'

9'

Height, ft

128

90

CHART III.   Soviet Space Boosters

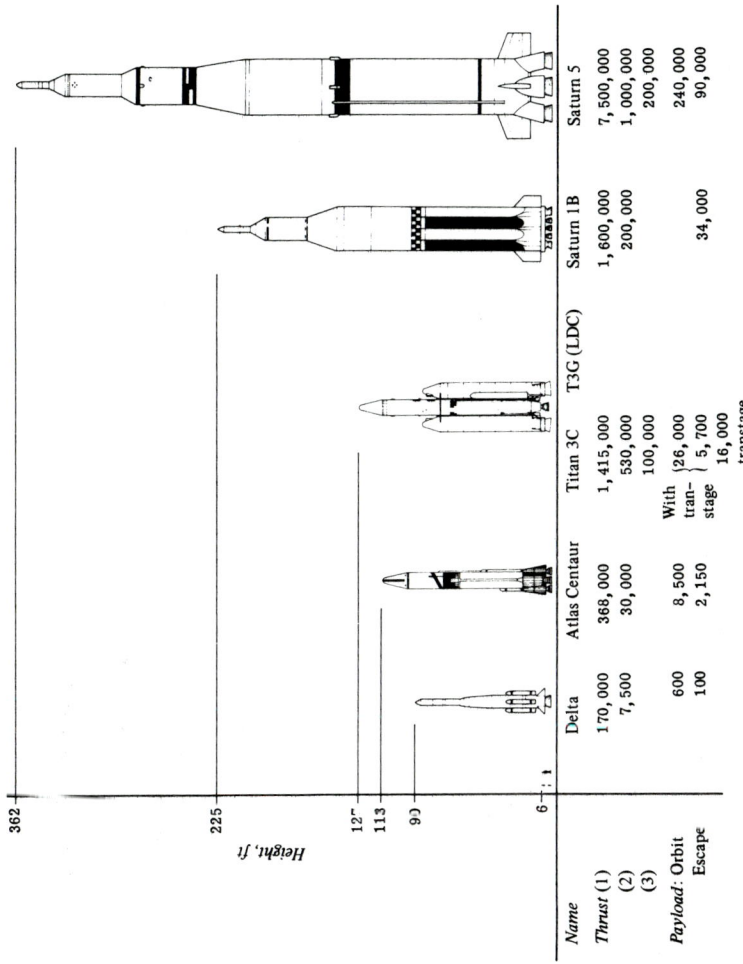

CHART IV. U.S. Space Boosters

Name	Delta	Atlas Centaur	Titan 3C	T3G (LDC)	Saturn 1B	Saturn 5
Thrust (1)	170,000	368,000	1,415,000		1,600,000	7,500,000
(2)	7,500	30,000	530,000		200,000	1,000,000
(3)			100,000			200,000
Payload: Orbit	600	8,500	With tran-stage $\begin{cases} 26,000 \\ 5,700 \\ 16,000 \\ \text{transtage} \end{cases}$		34,000	240,000
Escape	100	2,150				90,000

Height, ft

362
225
127
113
90
60

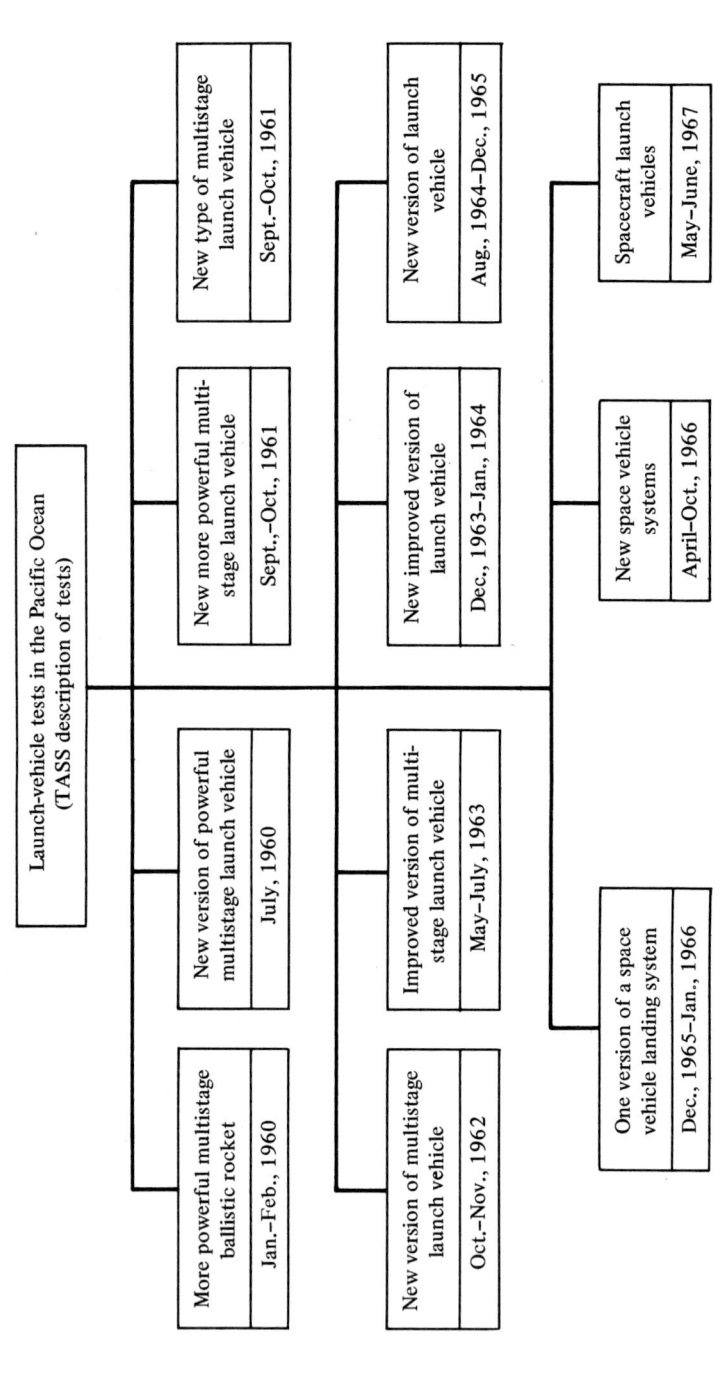

CHART V. Soviet Launch Vehicle Tests—Pacific Test Range

from which were derived the seven vehicles used thus far in the space program. Each launch vehicle apparently undergoes a familiar pattern of development: engine development, about five years; vehicle development, about three years; system tests, about one year; flight tests in the Pacific; product improvement; addition or improvement of upper stages.

Because some of these processes can overlap, the total process cycle, from engine development authorization to system test completion, is approximately seven years. The flight tests following this development cycle were compiled by the Library of Congress using the Tass description of the tests.

There is considerable evidence that all Soviet boosters from the first ICBM through the Vostok were designed by the same organization under the direction of the late Academician S. P. Korolyov. Recently, however, a second organization, as yet unidentified, was given the responsibility for the design and development of the Proton launch vehicle. It is the opinion of some analysts that the Proton design team has developed a launch vehicle that is totally unlike the configuration of the Vostok family. Responsibility for the development of all rocket engines, on the other hand, is still assigned to a single agency, the Leningrad Gas Dynamics Laboratory. Primary responsibility for the development of the Soviet lunar launch vehicle has not been identified in any press releases.

Spacecraft

Examination of the design of the Russian spacecraft reveals that an "add-on" philosophy similar to that used for the booster development was employed. This implies that special satellites are not developed for each mission; instead, the necessary equipment for specific missions are incorporated or added-on to an existing satellite. In this manner Soviet space research is able to carry out its program with less expenditure of time and money.

Sputnik 1 set the pace. It was a sphere 23 in. in diameter. The assembly system which stemmed from it began a year later with Luna 1. This satellite consisted of two air-tight-joined hemispheres 47 in. in diameter. Because a number of satellites were subsequently developed from this one, it is identified as the *Lunar genotype*.

As satellite sophistication and requirements increased, a 47-in. cylindrical section was added between the two hemispheres. The length of this section depended on mission requirements and varied from 24 in. to 47 in. In some cases, solar cells and/or sensors were added to the exterior. A number of Cosmos satellites used these configurations, as did the Electron series and the Luna 3.

The planetary series started with Venus 1 as a prototype. Venus 1 consisted essentially of a cylindrical body with attached solar-cell panels and a large parabolic antenna. Midcourse correction equipment and other apparatus not required for the planetary measurements were located in an auxiliary, smaller cylinder, joined axially to the body cylinder. Venus 1 is designated the *Planetary genotype*.

This spacecraft proved to be inadequate. After four launches it was modified by enlarging the body cylinder and the propulsion module and by adding hemispherical thermal radiator appendages. This second version proved more dependable and was used in subsequent planetary flights. Venus, Mars, and Zond satellites all use this configuration. Modifications of this vehicle resulted in a new series of spacecraft including the second-generation Luna series, the Molniya, and the Cosmos meteorological satellite.

A third series of assembly-type satellites began with Sputnik 4, later identified as Vostok. These large recoverable satellites consisted of the final stage of the carrier rocket, an equipment section, a retrorocket system, and the reentry sphere (comparable to our Mercury, Gemini, and Apollo in function but not in appearance).

The entire Vostok assemblage remains together in orbit until reentry is initiated. Then, the spherical capsule separates from the final stage, orients itself, and fires its retro system. After deceleration, the retro system is jettisoned and the spherical capsule reenters. At the appropriate time, the cosmonaut or instrument package is catapulted out of the sphere, and both are soft landed by parachute. This satellite has been very successful. It has been used for all biological payloads, including all manned flights, and is also used for the extensive Cosmos reconnaissance series. Elements of the Vostok also appears in the Soyuz spacecraft and station. This last assembly-type satellite is called the *Recoverable genotype*.

The last of the genotypes is the Proton E-1. Only one derivative of this type has been used.

To date, thirteen different names have been associated with the many different Soviet launches. Usually the word used to refer to the program is an adjective describing the mission; frequently the name remains as the program name.

Ground rules adopted by Soviet scientists to guide the development of early sputniks was carried over into the various categories of specialized spacecraft that followed. Their problems were compounded, however, because of the variety of tasks associated with space research, the wide range of applications, the numerous requirements imposed on the satellite by the planned experiments and the large number of scheduled launches. From a practical standpoint, these cumulative demands eliminated the possibility of designing a single multi-purpose satellite suitable for solving all research problems. To resolve the mission multiplicity problem, Soviet scientists adopted the spacecraft standardization concept. To make it work, their designers grouped all of their requirements into related research experiments, enabling minimal modifications to fulfill most of the mission and experiment requirements.

In practice, standardization permits the same spacecraft shell or body, the same service systems, on-board equipment, control circuitry, and power-supply systems to be adapted to the various mission needs. Thus, it was possible for them to go quickly from one mission to another, retaining maximum design continuity between spacecraft applications and modifications. In general, the basic structure of the Soviets' standardized satellite is a cylinder with hemispherical ends. The internal volume was divided into three functional compartments: research instrumentation; service equipment; and power supplies. The external surface is fitted with brackets and flanges for mounting sensors

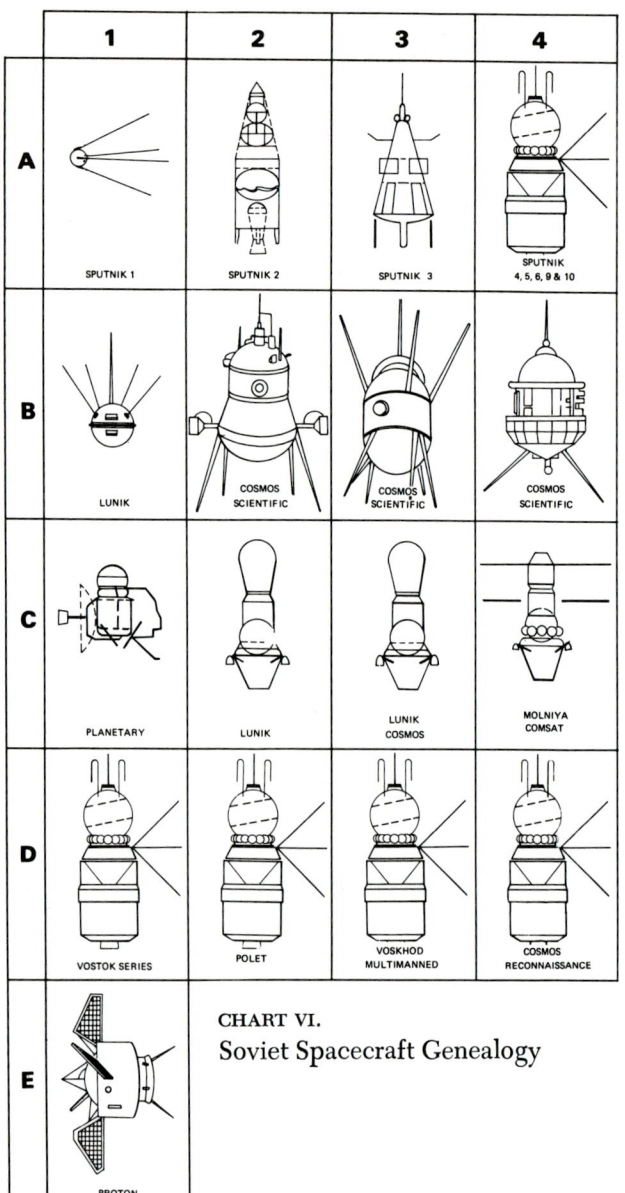

	1	2	3	4
A	SPUTNIK 1	SPUTNIK 2	SPUTNIK 3	SPUTNIK 4, 5, 6, 9 & 10
B	LUNIK	COSMOS SCIENTIFIC	COSMOS SCIENTIFIC	COSMOS SCIENTIFIC
C	PLANETARY	LUNIK	LUNIK COSMOS	MOLNIYA COMSAT
D	VOSTOK SERIES	POLET	VOSKHOD MULTIMANNED	COSMOS RECONNAISSANCE
E	PROTON			

CHART VI.
Soviet Spacecraft Genealogy

104

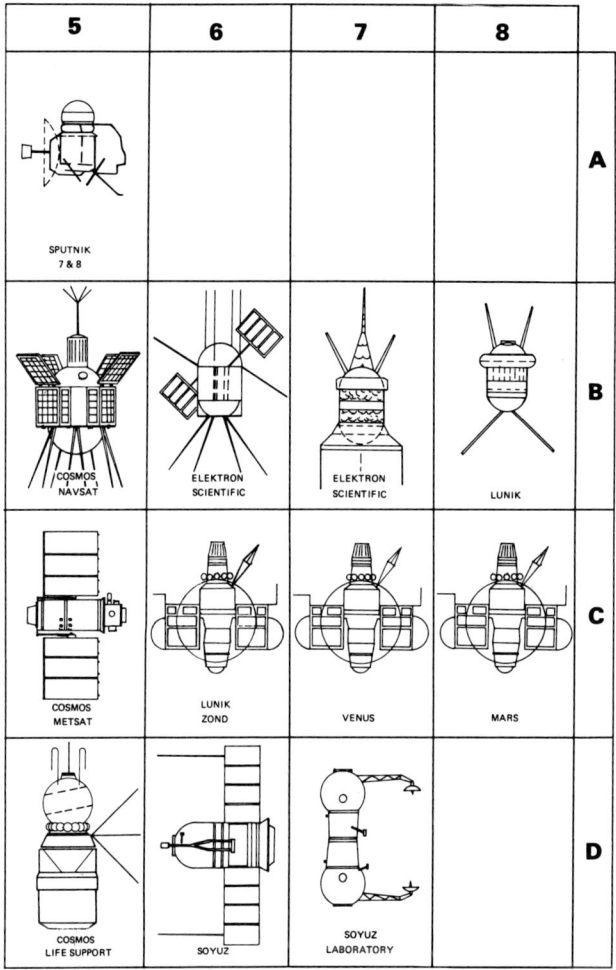

	5	6	7	8
A	SPUTNIK 7 & 8			
B	COSMOS NAVSAT	ELEKTRON SCIENTIFIC	ELEKTRON SCIENTIFIC	LUNIK
C	COSMOS METSAT	LUNIK ZOND	VENUS	MARS
D	COSMOS LIFE SUPPORT	SOYUZ	SOYUZ LABORATORY	

Within the few illustrated genotype spacecraft one can account for almost all of the Soviet space launchings from October 4, 1957, through December 31, 1969. For example, column 1, B through E inclusive, depicts the genotype of each spacecraft family. Whereas row A shows the highly publicized Sputnik launches; rows B, C and D show the derivatives of the Luna, Planetary, and Vostok series; row E shows the three launches of the Proton genotype only.

105

and other instrumentation directly on the body. The number of brackets and flanges and their sizes are chosen to provide adequate sealed leads for multiple applications.

In the standard satellite design, special attention was paid to the temperature regulating characteristics. It is usually heated by solar radiation and by the earth's reflected radiation, as well as by the heat released from its on-board equipment.

The heating of a satellite by solar radiation depends largely on the time it remains on the illuminated or sun side of the earth. Spacecraft in a low orbit are normally exposed to the sun for up to 70 percent of their orbital period. In special cases, an orbit may be such that a satellite will be exposed to the sun continuously for several days. The flux of solar radiation reflected by the earth depends on the nature and characteristic of the earth's surface, on the position of an orbit relative to the illuminated part of the earth's surface, and on the orbital position of the satellite. On the average, the earth's reflected flux amounts to approximately 35 percent of the flux of direct solar radiation.

Besides absorbing heat, a spacecraft radiates internal heat from its outside surface. The lower hemispherical end of the Soviet standardized satellite body is used as a radiator, which consists of a ceramic coating that characteristically has a high radiating capacity. The coating is deposited in sections on the lower outer surface of the hemisphere. To control the amount of heat that is permitted to radiate into space, shaped louvered shades, sized to match the ceramic sections, are moved to cover or expose the surface of the ceramic radiator, as required, to regulate the satellite's internal temperature. The remainder of the outer surface of the satellite is specially painted or polished to bring into proper balance the solar radiation absorbed and the heat radiated from the satellite.

To transfer the heat released from the equipment within the satellite's structural shell, the internal volume of the satellite is filled with an inert gas, such as nitrogen. A fan is then used to circulate the gas, which carries the heat from the instruments to the walls of the shell. Free gas convection does not occur in a "weightless" condition.

To make the standardization concept applicable to the majority of missions, Soviet designers were required to extend the commonality factor beyond the spacecraft configuration, structure, and thermal control systems to include the spacecraft's orbital orientation systems, recovery system, and onboard auxiliary equipment. Some space missions require that the spacecraft be oriented for a long time with a predetermined accuracy with respect to the earth, sun, or stars. In each of these cases, it is essential that special sensors and instrumentation be oriented in a known or specific manner with respect to the spacecraft and the celestial body under observation being used as reference.

To obtain or maintain the necessary earth or celestial orientation and long-term stabilization, two standard techniques were used. They were microrocket motors and momentum wheels rotating inside the satellite. In most cases, these methods were not used by themselves but in conjunction with the other device.

Sun or celestial orientation for their satellites is obtained by combining the microrocket motors with an internal flywheel system. Normally, the micromotors balance out spacecraft disturbances caused by separation from the orbital injection stage. The flywheel, activated by solar sensor signals, is used to orient the spacecraft with the celestial object under observation. In the event that the flywheel introduces extraneous disturbances during the orientation phase, the micromotors are activated again to neutralize these secondary effects. In essence, the micromotors are used for extremely fine trim of the spacecraft.

Perhaps one of the most important provisions that was incorporated into the spacecraft standardization concept was the provision for a recoverable capsule. Recoverable capsules are essential in cases where the scientific experiments or experimental objects must be returned to earth for final analysis and evaluation. This capability is also necessary for military missions, such as surveillance and reconnaissance.

In the Soviets' standard satellite design, the nonrecoverable and recoverable versions are identical in all respects but one:

in the recoverable version, the research instrumentation and experimental objects are packaged in a special autonomous compartment or capsule that is recovered after completion of the mission. By this approach, the standardization concept is maintained, because the packaging of the research instrumentation in the capsule does not involve any change in design.

In orbit a satellite moves at a speed of more than 25,000 ft per sec relative to the earth. In the recovery phase, the velocity of the returning satellite must be reduced to less than 25 ft per sec on landing. At the present time, aerodynamic drag and braking motors are employed by both the United States and the Soviet Union for decelerating satellites. The recoverable portion of the spacecraft is supplied with a special rocket-braking motor that is ignited during the final seconds before touchdown. The spacecraft is also provided with special drag devices and parachutes to greatly decrease the rate of the satellite's descent from about 100,000 ft altitude to touchdown.

The on-board auxiliary equipment in the Soviets' standardized spacecraft concept has been developed for operation in conditions of weightlessness and extremely high launching acceleration. The characteristics of this equipment, according to Soviet scientists, are long-term operational reliability, light weight, small size, and minimum power consumption. According to the available launch statistics, reliability, except for interplanetary spacecraft, has been good.

Information gathered by the standard satellite is transmitted to earth via a multichannel, high information-rate telemetry system. On-board telemetry equipment transforms the scientific measurements into radio signals of a specific modulation form. These signals are transmitted to ground stations where they are restored to their original form for analyses or study.

Information can be telemetered to earth only when the satellite is within line-of-sight and communications range of the ground station. This necessitates that most of the sensor information developed during the orbital period be stored on tape or film for later read-out. This approach is used by both the United States and the Soviet Union. Since the tape playback speed can be increased, it is possible to read out complete orbit information (approximately 90 minutes' worth) in a matter of 5 minutes.

In most operations, the scientific instruments carry out experiments at many points in the orbit under a predetermined program. The information-storage system collects this data and retains it for one or more orbits, depending on the mission and operation. During the ground-to-satellite-to-ground communication, scientific experimental data, spacecraft operational data, and satellite positional data are directly transmitted.

The on-board equipment in the standardized satellite is normally controlled by a ground-to-satellite-link command radio or by an on-board timing device. It is the ground-command control center however that controls the flight of all satellites. It measures the parameters of the orbit, receives and records telemetry data, and sends commands to switch the satellite-borne instrumentation on and off as required.

In general, the design of the standardized satellite permits the Soviet experimenters to go from one experiment to another and from one mission to another by simply making adaptations to the standard spacecraft. But standardization in any form is usually the result of compromises and inefficiencies and, moreover, standardization is usually more difficult to achieve in practice than in theory.

9 Earth Satellites

By settling around earth in a multitude of rings similar to the rings of Saturn, people will insure the supply of solar energy by a factor of 100–1000 compared to what they have on the surface of earth. However, this may not satisfy man either, and having conquered these bases, he may extend his hands for the remaining solar energy which is 2 billion times greater than that received by earth. In this case, eternal motion around the earth must be replaced by similar motion around the sun. For this purpose, it will be necessary to move farther away from earth and become an independent planet—a satellite of the sun and a brother of earth.

K. E. Tsiolkovsky, 1911

THE MOST complex series of earth satellites developed by the Soviet is the Cosmos. Cosmos actually is a cover name given to a very large variety of space launches. At times, the name is used to obscure the details of both successful and unsuccessful flights, as well as a large family of scientific and perhaps military satellites. In this regard, the Soviets have never revealed the weight of a Cosmos satellite. Although recently they have publicly displayed some of the Cosmos satellites without identifying the types associated with specific missions, it was obvious that the Cosmos payload-weight categories covered a wide range.

The first of the Cosmos satellites was launched on March

16, 1962. By December 23, 1969, 317 Cosmos-type satellites were put into orbit around the earth. The orbital parameters, size, and weight of these satellites varied widely according to their orbital mission.

On the day that the first Cosmos was launched the Soviets also outlined the goals of this new program: the exploration of the upper atmosphere and ionosphere; the study of cloud systems in the earth's atmosphere, to facilitate the development of components for a meteorological satellite system; the measurement of charged particles in the ionosphere and radiation belts; the measurement of cosmic rays and corpuscular fluxes; the study of the earth's magnetic field; and the measurement of short-wave emissions from celestial bodies.

Cosmos Satellites

Military Communication Satellite   Between August, 1964, and September, 1965, approximately twenty-four satellites were launched in groups of three and five from Tyuratam. These launches have not been specifically identified nor have any

FIG. 36   Cosmos radiation measurement satellite.

FIG. 37   Cosmos navigational satellite.

other multiple launches taken place since that time. In the United States multiple launches are used primarily by the interim defense communication satellite system. In the initial U.S. launch, the first seven satellites of a 22 communication satellite system were successfully placed in a synchronous orbit.

Navigation Satellite (*Navsat*)   Between May, 1967, and December, 1967, a third type of Cosmos satellite was launched from Plesetsk at an inclination of 75 degrees. This group of satellites have tentatively been identified as navigational type satellites, although there has been no official confirmation by either the Soviet Union or the United States.

Life Support Satellite   Perhaps one of the more unique applications of a Cosmos satellite was Cosmos 110, which was the third satellite used in the Soyuz manned-spacecraft test flights. For this flight, a Vostok spacecraft was modified to test a new long-term (sufficient for a lunar flight and more) environmental life-support system. The test subjects in this case were the dogs, Veterok and Ugolek. The flight began on February 22, 1966, and continued for twenty-three days, after which the dogs were successfully recovered. It is presumed that this test was a success, since no other flights of this type followed.

Reconnaissance Satellites Probably the largest number of Cosmos-type satellites have been launched for military reconnaissance missions. Cosmos 4, launched on April 26, 1962, was the first of the series. It remained in orbit three days, and its booster remained in orbit fifty-two days. The recovery of Cosmos 4 was conducted without comment from the Soviets.

Three months later, Cosmos 7 flew in a similar orbit and stayed in orbit four days. Since then more than 140 different Cosmos launches have gone into similar orbits. Whereas the early reconnaissance launches were from Tyuratam and remained aloft for eight days, the more recent launches have

FIG. 38 Cosmos space environment satellite.

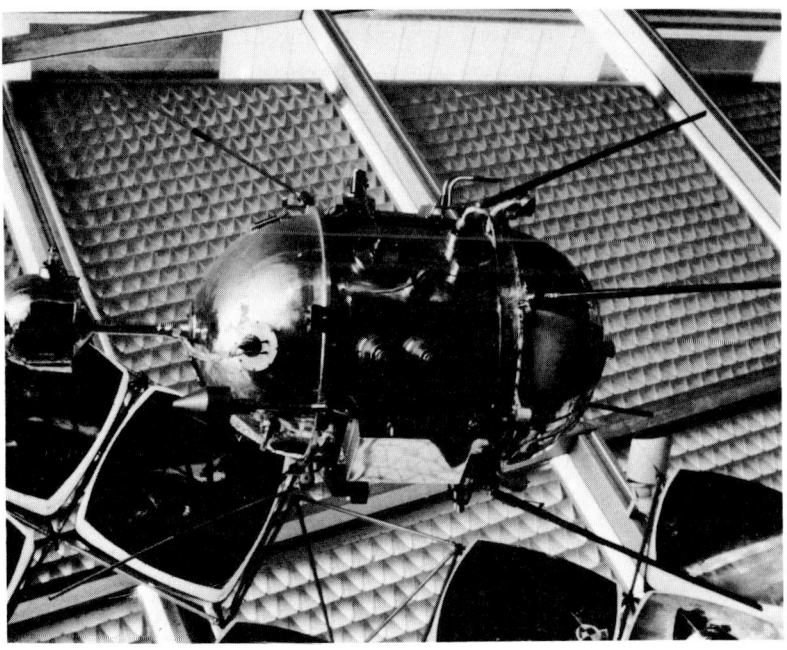

been made from Plesetsk and stay in orbit for about twelve days. The increase in flight duration and switch to Plesetsk strongly indicates that the reconnaissance satellite has now achieved operational status.

Both the United States and the Soviets use their reconnaissance satellites to identify major changes in military establishments, to count missiles and bombers on the ground, and to observe naval vessels under construction in shipyards. In 1969, the Soviet Union launched the record number of thirty-two observation satellites to monitor both the United States and China from space.

Fractional Orbital Bomb System (FOBS)   Since September, 1966, fourteen flights have been identified that have had similar characteristics, i.e., they all completed their mission within one orbit and were returned to earth on command. These flights have been interspersed at random between Cosmos 139 and Cosmos 298. A fractional orbital bombardment system is a flat trajectory missile which can go around the world eluding the American warning radar system in Greenland.

Between 1966 and 1967, the Soviets conducted a fast paced FOBS flight test program, firing eleven of the vehicles by the end of 1967. However, the launches dwindled to two in 1968 and only one in 1969.

One interpretation of reduction in FOBS launchings is that the Soviets have completed their development testing and are ready to deploy the weapon. David Packard, Deputy Secretary of Defense, cited the FOBS threat in arguing for Senate approval of the Safeguard ABM defense.

Maneuvering Satellite   Another highly speculative group of satellites which made a dramatic appearance and then just as suddenly were drastically curtailed are the mysterious maneuvering satellites. The first of this series began with Cosmos 198, launched on December 27, 1967. This was quickly followed by five launches in 1968 and then decreased to one in 1969.

Just what this satellite mission is supposed to be is still unknown to American weapon specialists. One theory is that the

Soviets were developing a satellite killer, sending up a target satellite and then exploding another near it. This theory is in part substantiated by the many pieces of space vehicles which have been detected by radar, indicating that an explosion of some kind had taken place.

Unmanned Rendezvous The Soviet flight of the first unmanned rendezvous and docking mission was accomplished by Cosmos 188 and Cosmos 186. Cosmos 186, which was sent into orbit on October 27, 1967, was the active satellite. During the next three days, the on-board systems were checked. Orbital maneuvers were performed on October 29 so that the satellite would pass over the launch point on October 30. At the opening of the launch window on October 30, 1967, Cosmos 188 was sent into orbit.

The distance between the two satellites at the moment of the launching was approximately 15 miles, and the relative velocity between the two Cosmos spacecraft was about 56 mph. After injection of Cosmos 188, the two spacecraft had to locate, maneuver, approach each other in space, and then dock. To perform this mission both satellites were equipped with special targeting equipment and docking units. All operations were controlled by on-board communication and computing equipment, and the mutual search, approach, and docking in space was done automatically.

To accomplish the docking, Soviet engineers provided one of the satellites with an active docking subassembly, or "bar." The other satellite carried a passive docking subassembly, or "receiving case." The receiving case with the docking seat was the target into which the bar entered in the final stages of the satellite rendezvous.

On the forty-ninth orbit of Cosmos 186 and the first orbit of Cosmos 188, the docking phase of the flight was completed. The actual docking took place outside the line of sight of the Soviet Union. A photographic record was made of the entire docking process. When the docked satellites came within range of the Soviet tracking station, television pictures of the docked satellites were transmitted to earth. After 3.5 hours in a docked position, the satellites were decoupled by a radio command from earth, and propulsion systems aboard the satellites then transferred the satellites into different orbits.

Rendezvous and docking can be accomplished either with manned or unmanned spacecraft. In the first decade of the space age, the United States' Gemini program demonstrated repeatedly man's capability to conduct this type of maneuver in orbit. In the second decade of the space age the United States' Apollo program again demonstrated this capability in both earth and lunar orbits. The Soviet Union, on the other hand, demonstrated the feasibility of accomplishing rendezvous and docking with unmanned spacecraft and with the manned Soyuz 4 and Soyuz 5 in January, 1969. It is believed that the Soviets were trying to achieve an orbital Troika rendezvous and docking, in October, 1969, with the launches of Soyuz 6, 7, and 8. There were no confirmed reports that docking was attempted or accomplished. In the final analysis, rendezvous and docking either manned or unmanned, is indispensable to space flight, and nations engaged in space flight must eventually develop and use both methods on a day-to-day basis.

Elektron Satellite

In special cases where the scientific mission requirements could not be fulfilled by the Cosmos family of spacecraft, Soviet scientists developed special-mission-oriented satellites, in this case the Elektron and the Proton. Surprisingly, even in these specialized cases they made maximum use of standardized components.

The Elektron was developed to explore the Van Allen belt. Soviet scientists believed that the radiation belt was an exceptionally dynamic phenomena associated with other geophysical and solar activities. Accordingly, they decided to carry out simultaneous measurements of the belt at different points in near-space. This required a special space system of at least two satellites placed in elliptical orbits so that the upper atmosphere, the earth's radiation belt, and near-space could be investigated at the same time.

FIG. 39 Elektron 1 and 2, launched together on January 30, 1964.

On January 30, 1964, the first dual satellite system, consisting of Elektron 1 and Elektron 2, was successfully orbited. Elektron 1 had an apogee of 4,409 miles and a perigee of 252 miles, whereas Elektron 2 had an apogee of 42,352 miles and a perigee of 285 miles. Six months later on July 10, 1964, the second dual system, Elektron 3 and Elektron 4, followed the first series into orbit. Elektron 3 had an apogee of 7,040 miles and a perigee of 405 miles. Its mate, Elektron 4, had an apogee of 66,235 miles and a perigee of 459 miles. These were the last of the Elektron type satellite launches through December, 1969.

It is a well-known fact that earth satellites moving in low orbits are short-lived, as they are decelerated by the drag of the upper atmosphere. Normally, the higher the orbit the longer the satellite life, since the effect of the atmospheric drag decreases. For the perigee altitudes of the Elektron satellites, the drag effects of the atmosphere can be neglected. However, as the apogee altitude is increased to 18,000 or 25,000 miles satellite motion is decidedly affected by the perturbations of lunar and solar attraction forces. It was estimated that under the most unfavorable conditions, the life of a satellite whose apogee is around 40,000 miles would be only a few days. In this regard extensive studies were performed on satellites with high apogees. The launching altitudes were selected so as to ensure an adequate long life for these satellites.

The satellite orbits were inclined 61°. Like the altitude, the perigee at this inclination is affected by the moon, sun, and oblateness of the earth. In this case, the perturbances shifted the perigee of the orbit in a northward direction, which brought the satellite's orbital path through the thickest portion of the radiation belt within a year.

Having the orbital perigees in the northern hemisphere, in proximity to Soviet ground stations, resulted in favorable space to ground communication periods. It had an added significance because the volume of scientific information was the largest when the probes were in the perigee region. At this point, top-atmospheric measurements were made simultaneously with investigations of the radiation belts.

The obvious difference in the satellites was in the mounting of the solar cells. Elektron 1 utilized large, foldable, solar-cell arrays that were stored during the launch phase and deployed only after achieving orbit. Elektron 2, on the other hand, had its solar cells mounted directly on the body of the spacecraft.

The satellites had similar experimental functions. This included investigation of the radiation belts, measurement of the magnetic fields, detection of cosmic rays, analysis of the chemical composition of near-space, monitoring shortwave solar radiation and cosmic radio emission, and micrometeorite detection. Consequently, the instrumentation of the probes was almost identical.

There are four basic instruments that are used to measure the electrons and protons found in the radiation belts around the earth. These are a Geiger-Mueller counter, scintillator detector, mass spectrometer, and a so-called ion trap. All four of these instruments were included in each of the four Elektron satellites. They measured particle energies (electrons and protons) from 2 million electron volts (ev) down to less than 100 ev.

By definition, an electron volt represents the energy gained by an electron when it is accelerated by means of a voltage difference of 1 volt. Thus, 1 electron accelerated by a 100-volt potential will acquire 100 electron volts (ev); through 1,000-volts potential it will acquire 1,000 ev or 1 kev; through 1 million volts potential it will acquire 1,000,000 ev or 1 Mev.

As large as these numbers may seem, the actual energy of an ev is very small. In fact, 1 Mev can only light a 100-watt bulb for 0.032 trillionths of a second, a time span not possible to measure with present-day technology.

The motion of the particles in the radiation belt is determined by the earth's magnetic field. Therefore, a survey of the radiation belts was made in conjunction with measurements of the field. To make these measurements, two magnetometers were installed on both Elektron 2 and Elektron 4.

Under the influence of a multitude of factors, many of which are still unknown, particles spill out of the radiation belt and bombard the upper atmosphere. Thus, to a certain extent, the radiation belt affects the earth's atmosphere. Because of this interrelationship, the composition of the upper atmosphere was investigated simultaneously with the radiation belt. To conduct this particular experiment, Elektrons 1 through 4 carried radio-frequency mass spectrometers to determine the ionic chemical composition of the upper atmosphere.

In addition to gas molecules, atoms, electrons, and protons, there are a large number of dust particles in the earth's vicinity, known as micrometeorites. Earlier American and Soviet satellites ascertained that the concentration of micrometeoric particles in the neighborhood of the earth is greater than in interplanetary space. The investigation of meteoric bodies is important for understanding what part they play in the upper atmosphere, particularly in the formation of noctilucent clouds, atmospheric luminescence, and other phenomena.

To measure the meteoric particles in space and determine their energies and masses, Elektron 1 carried a Piezo electric sensor. In this instrument, electric signals generated by the pressure created by the impact of a micrometeorite are converted and sorted according to the force of the impact. The sorted impacts are then divided into several groups containing the number in each impact group.

During solar flares, powerful X rays are emitted by the sun, which strongly affects the earth's ionosphere, particularly the D layer of the atmosphere. To monitor the solar flares, Elektron 1 and Elektron 3 carried instruments to measure solar X-ray emissions. In contrast, during a period of minimum solar activity, lengthy measurements of the sun's X-ray emission and chemical composition of the outer ionosphere (up to altitudes of 1,860 miles) were measured with Elektron 2 and Elektron 4. Ions of hydrogen, helium, nitrogen, and oxygen were detected.

Just as X rays are emitted from the sun, streams of charged particles of immense energies, known as *cosmic rays,* come to the vicinity of the earth from extragalactic space. The earth's atmosphere and magnetic field impede them considerably. The magnetic field deflects some cosmic rays away from the neighborhood of the earth while others are mashed in collisions with gas molecules surrounding the earth, reaching the earth's surface only as low-energy particles.

The Elektron 2 and Elektron 4 in their orbits traveled beyond the atmosphere and the geomagnetic field. They carried instruments to measure the over-all intensity of cosmic rays. They also determined the physical composition of cosmic rays.

Elektron 1, like many of its predecessors, carried a radio transmitter for investigating the structure of the ionosphere and interplanetary space with the aid of radio waves in the frequency range of 20.005, 30.0075, and 90.0225 *mc* per sec. In the lower frequency range, the radio waves are reflected off the ionosphere. In the higher frequency ranges, that is 100 *mc* per sec or greater, the radio waves penetrate the ionosphere and travel out into space. Radio-transmission experiments for investigating the ionosphere are included in many of the Soviet experimental satellites.

Proton Satellite

In the stream of cosmic rays showering the earth from outer space are particles of different and extremely high energy levels. One of the key problems in the study of cosmic-ray physics is the question of how atomic nuclei of different elements acquire such high energies. For the accurate quantitative measurements of cosmic-ray particles with high and superhigh energies, Soviet scientists designed equipment that automatically classified these particles by energies, selecting those which possessed very high energies and then determined the nature of the particle. To perform this experiment on a satellite, it was estimated that a spacecraft exceeding 20,000 lb in weight was needed.

The Proton 1 satellite, launched into orbit on July 16, 1965, was built specifically to study superhigh-energy cosmic-ray physics. The satellites' construction is basically a cylinder within a cylinder. The internal sealed body protects the satellite against aerodynamic forces and temperature effects while it is being boosted into orbit. To prevent overheating and cooling in orbital flight, the body of the satellite is coated ex-

FIG. 40   Proton satellite.

122

ternally with a thermal insulator. Inside the cylinder, in the rear and central sections, research instrumentation and auxiliary systems are located. The body of the satellite is sealed with dome-shaped ends. Normal atmospheric pressure and operating temperatures are maintained within the instrumented satellite structure.

The Proton's primary source of electrical power is from solar cells. Solar-cell banks are mounted externally on special panels which are folded in the shape of a truncated pyramid before the spacecraft is launched into orbit. In orbit the panels unfold and are fixed in position forming something like a four-bladed propeller. The solar cells power the instrumentation on the sunny side of the orbit and charged batteries supply the power on the shaded side of the orbit. The batteries are located between the outer structural shell and the internal, cylindrical instrument container. Special sensors, to keep the solar paddles oriented toward the sun, are also mounted externally. These sensors work in conjunction with a compressed-gas system and gas nozzles. The gas nozzles, control equipment, and the external thermal-control system are installed on the rear of the satellite, as in all standard systems. Other externally mounted equipment includes the telemetry antenna, command, and trajectory monitoring equipment. The telemetry system is the standardized version.

The research instruments of the Proton space station include an ionization calorimeter for measuring the energy of cosmic-ray particles of high and superhigh energies and investigating the characteristics of their interaction with matter. The calorimeter consists of a large number of steel plates with scintillators made of special plastics between them. Two special counters are installed above to measure the electric charge of cosmic-ray particles. Each counter measures it independently. The application of two counters substantially increases the accuracy of charge measurements and makes it possible to separate primary high-energy protons from heavier particles.

Under the counter there is a plate of carbon in one half of the instrument and a plate of polyethylene in the other. Interaction detectors are installed underneath the plates. Polyethylene consists of carbon and hydrogen molecules. Therefore, in one half of the instrument there are interactions of ultra high-energy particles with nuclei of carbon ions and in the other with those of carbon and hydrogen ions.

When a high-energy particle hits the ionization calorimeter, the collisions produce secondary particles that, also colliding with ion nuclei, produce particles of the next generation, and so forth. As a result, the entire energy of primaries passes to a large number of secondaries, which are absorbed into the substance of the ionization calorimeter. Comparison of the measurements carried out in the two halves makes it possible to isolate the interactions with nuclei of hydrogen ions (protons).

The absorption of energy is accompanied by flashes of light in the plastic scintillators, the intensity of each flash being proportional to the energy absorbed. Installed under the ionization calorimeter is a scintillation counter that, in combination with the interaction detector, detects particles moving in a definite direction.

The scientific investigations of the superhigh-energy elementary particles carried out by the Proton 1 were continued by Proton 2, placed in orbit on November 2, 1965, Proton 3, orbited on July 6, 1966, and Proton 4 orbited November 16, 1968. All four satellites were comparable. This was the last of the Proton launches through December, 1969.

Application Satellites

Included in the over-all Soviet orbital space program are a number of missions directed toward performing functions or services that are now carried on by earthbound systems. Satellites used on these missions are generally referred to as "Earth Application Satellites."

It is not expected that the application satellite systems will completely replace earthbound systems but will complement those systems. Currently the primary emphasis in both the United States and the Soviet Union is on communications and meteorological satellite systems.

FIG. 41 Molniya communications satellite, launched April 23, 1965
—weight approximately 1,800 lb.

Molniya Communication Satellite (Comsat) Molniya is the
name given by the Soviets to their communication satellite. Its
design appears to have been adapted from the planetary family
of spacecraft. Its major distinction seems to be a significant
modification of the solar-cell paddle configuration. The basic
function of the Molniya satellite is to transmit telegraph, telephone, radio, and television information. This is, in effect, a
television tower some 24,840 miles above the earth.

Prior to the Molniya operational flights, two precursor test
vehicles were launched. Both of these flights ended in failures.
The first flight, Cosmos 21, failed during an attempted transition from a circular orbit to a highly elliptical orbit. The second failure, Cosmos 41, was a test of a complete operational
system.

The first launch of a successful operational system occurred
on April 23, 1965. Since then, a total of eight launches have
been made with only one failure. Had it been a success, it
would have been designated Molniya 1F. Instead, it was given

the cover name of Cosmos 174. It is interesting to note that instead of the usual numbering system (Molniya 1, Molniya 2, etc.) these launches were designated Molniya 1A through Molniya 1G. One possible explanation is that a second generation of the Comsat-type satellite is planned and that these will be called the Molniya 2 series.

The Molniya family of satellites are equipped with solar cells, a radio relay system and parabolic antenna, and an orbital-altitude-control system. The latter is comprised of an orbital-altitude-correction engine, optical orientation sensors used in conjunction with gas jets for orbital control, and a gyroscopic altitude-sensing and stabilization system.

The satellite consumes much of the electrical power it derives from the solar-cells power supply. For purposes of compact packaging and deployment, the solar cells are mounted on six movable paddles that open to their operating positions on attaining orbit.

The proper orbit is assured through the operation of the orbital-altitude-correction engine. On attaining the proper orbit, the banks of solar cells are oriented toward the sun through the combined action of the satellite's optical sensors and altitude-correction and stabilization systems.

The ends of the Molniya parabolic antennas spring up perpendicularly to the solar paddles and open like umbrellas. The antenna must be accurately pointed so that all of the propagated radio energy is directed toward the earth. This is achieved through the use of an optical earth sensor, which is mounted on the antenna. If the antenna fails to operate, a standby antenna takes over the transmission task. All on-board instruments operate on command from earth.

The satellite has a special outer casing as an additional protection in flight. The temperature inside the communication satellite is automatically regulated by a temperature-control system. The satellite's high-altitude orbit causes it to travel for lengthy periods through the earth's radiation belt. Radioactive particles incessantly bombard its radio-relay equipment, which can affect its performance, even putting it out of operation. The satellite is, therefore, equipped with radiation dosimeters to measure levels of radiation, so as to study the effects of radiation on all spacecraft systems.

The apogee for all Molniya satellites occurs above the Northern Hemisphere with a period of about 12 hr per rev. This assures the reception of TV broadcasts over most of the northern sector of the globe.

With the apogee in the Northern Hemisphere, communication between different areas is possible for approximately nine hours during each revolution. During the first revolution, communication is possible between all points in the Soviet Union, and between there and many European and Asian countries. The second orbit makes possible a communication link between Europe, the Soviet Union, and America. The launching of two or more satellites into different orbits makes it possible to achieve round-the-clock communications with all points of the Soviet Union. The satellite's 40-watt radio equipment must receive and transmit many simultaneous telephone and telegraph channels. The satellite radio equipment receives transmissions from earth and relays them back. During the operation, there is a continuous telemetric check on all systems. In an emergency, standby equipment is switched on.

At first, the satellite was used for experimental transmissions, which included radio-telephone conversations, telephotographs, telegrams, and experimental color-TV transmissions. At the same time, work was started on direct TV communication between Moscow and Paris. On November 19, 1965, the first color-TV transmission between the two countries was made. For the test, Moscow transmitted several scenes and cartoon films. The Russian Molniya and the French SECAM satellites supplemented each other. The tests were successful, and work between the two nations is continuing.

After a series of experiments the commercial Moscow–Vladivostok long-distance telephone communication line was put into operation. Subsequently the use of the satellite link was also used for the exchange of television broadcasts between Moscow and Vladivostok.

On May 18, 1966, one of the Molniya 1 communication satellites took TV photographs of the earth from a distance of almost 24,840 miles. The TV cameras were installed on the outside of the spacecraft and were equipped with variable lenses and color filters. The photographs from this altitude made it possible to observe weather conditions over vast areas and to spot regions covered by cyclones and anticyclones, proving that Comsats of the Molniya 1 type can be used for meteorological forecasting.

The Soviets found that the orbital communication system was more economical than ground based relay network, particularly if the transmission was carried over natural obstacles such as oceans, mountains, or the tundra of the far north. Satellite communications involve only three elements: transmitter, satellite, and receiver.

The Soviets have replenished their communication satellite system on a yearly basis, launching Molniya 1-H, 1-J, and 1-K in 1968 and Molniya 1-L and 1-M in 1969.

FIG. 42   Cosmos meteorological satellite.

FIG. 43 Cosmos 122, precursor to meteorological satellite, launched June 25, 1966.

Meteor-Weather Satellites (Metsat) The Soviet meteorological satellite (Metsat) was first revealed at the Paris Air Show as a member of the Cosmos family. A study of the Soviet records indicates that the first Metsat launch probably occurred on December 11, 1963, with the orbiting of Cosmos 14. The records also indicate that by the end of 1967, fourteen meteorological systems (experimental and operational) had been placed into earth orbit.

The meteorological satellite is somewhat larger than the average Cosmos satellite and is made up of three cylindrical sections with two large solar paddles. It is about 4 ft in diameter, 12 ft long, and 18 ft from tip to tip of the solar panels.

The first step in the development of a weather satellite system consisted of launching Cosmos satellites equipped to test the electromechanical equipment designed to stabilize and orient the satellite toward earth. It was found that prolonged flight and continuous operation resulted in a deterioration of the solar cells. As a result, a better cell was developed. In the

129

experimental phase of development, it appears that eight Cosmos satellites were used, Nos. 14, 44, 45, 58, 65, 92, 100, and 118.

In the second stage of development, Cosmos 122 was test flown on June 25, 1966. This satellite carried two independent orientation systems: one was for the continuous pointing of the instrument compartment toward earth; the other was for orienting the solar panels toward the sun. The equipment designed to accomplish this used special photoelements which sensed only predetermined regions of the electromagnetic spectrum. For determining the direction to earth, instruments were used that reacted to the earth's thermal radiation. Since these units were very sensitive to electromagnetic disturbances, steps were taken to eliminate the influence of these.

Cosmos 122 yielded extensive meteorological information. In addition, the satellite carried actinometric equipment to obtain data concerning the radiation balance of the earth-atmosphere system. Cosmos 122 transmitted meteorological data for a period of four months.

The third step was the launching of Cosmos 144 and Cosmos 156 on February 28, 1967, and April 27, 1967. These satellites formed the experimental Metsat-meteor system. The relative positions of the orbits of the satellites were chosen so that they would produce weather observations above each of the earth's regions within a time span of six hours. The two satellite systems make it possible to obtain complete information from half of the earth's sphere within a few days.

Cosmos 122, 144, 156, and 184 are the Soviet Union's first long-lived meteorological satellites. These satellites employ a three-axis orientation system utilizing momentum wheels. The satellite operates continuously for a one-year period. The primary source of power is supplied by solar panels which are used in a sun search-and-lock mode. Chemical batteries are also carried for secondary applications.

The on-board instrumentation makes it possible to photograph clouds, snow cover, and ice fields, both in the dark and sunlit parts of the earth, and to measure radiation fluxes reflected and emitted by the earth-atmosphere system. Televi-

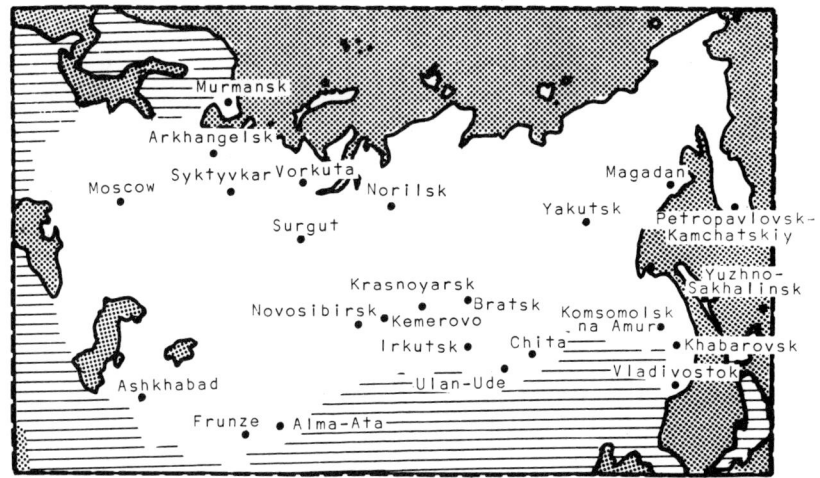

FIG. 44 *Top:* Location of receiver stations in orbital network. *Bottom:* Diagram of the stations: 1) feeder wave-guide channel, 2) activator, 3) guide system, 4) coupler, 5) reception monitor, 6) low-noise input system, 7) receiver, 8) system for separating image and sound, 9) audio monitor.

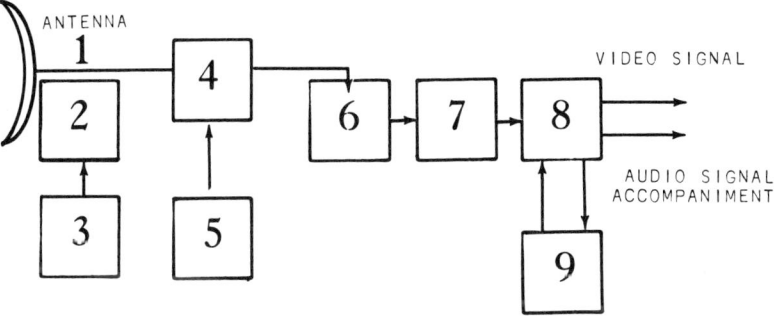

sion cameras are used to observe cloud formations on the sunlit sides of the earth, and two cameras continuously photograph the earth's surface along the satellite's trajectory. Observations of the night-side of the planet are made by an infrared televisionlike instrument. The television equipment is switched on automatically whenever the angle of elevation of the sun over the horizon exceeds 5°. At an altitude of 388 miles, the television cameras can photograph a strip of earth approximately 621 miles wide along the orbital flight path.

131

Table 2
Soviet Spacecraft Inventory Status, 1967

Name	CLASSIFICATION														Disposition	Projection
	Earth orbital	Lunar	Planetary	Scientific probes	Technology development	Comsat	Metsat	Navsat	Reconnaissance	Biological, unmanned	Biological, manned	Nonmilitary	Military	Average launches/year		
Elektron	X			X									X	?	Unknown, all still in orbit	May be integrated into multipurpose satellite
Cosmos	X			X									X	6	Continuing	May be integrated into multipurpose satellite
Cosmos	X								X				X	3	Replacement launches continuing	May be integrated into multipurpose satellite
Cosmos	X						X					X	X	3	Replacement launches continuing	May be integrated into multipurpose satellite
Cosmos	X							X				X		15	Continuing	Probable system sophistication leading to manned system

Name							No.	Status	Future
Cosmos	X	X			X	X		Test objectives probably accomplished	None—See Soyuz
Cosmos	X	X				X	X 11	Further tests a function of development status	Probable precursor to an operational military FOB system
Luna	X	X		X		X	3	Initial objectives probably accomplished	May be reactivated during manned exploration
Mars		X	X		X X		1	Problems solved, multiple launches each window	All landers, gradual technological buildup for manned expedition
Molniya	X		X	X		X	2	Replacement launches continuing	Candidate for multipurpose satellite or space station
Polyot (Polet)	X	X		X		X		Test objectives probably accomplished	None—See Soyuz
Proton	X	X		X		X	?	Probable first use of a multipurpose satellite	Series may be expanded in application
Soyuz	X	X		X		X	1	1 launch started new series	Probable prototype for large scale orbital & lunar operations
Sputnik	X		X		X	X	3	"Sputnik" terminology discontinued	None
Venus		X	X		X	X	2	Multiple launches expected each window	All landers, gradual buildup for possible manned expedition
Voskhod	X		X	X	X	X	2	Test objectives probably accomplished	None—See Soyuz
Vostok	X	X		X	X	X	3	Test objectives probably accomplished	None—See Soyuz
Zond	X X X		X	X		X	3	Continuing	Expanded application

The resolution of the photographs makes it possible to determine the shape of the clouds and analyze in detail the changing atmospheric conditions in a given specific area.

The results of the information obtained from meteorological flights will make it possible to develop a system of weather satellites in continuous communication with a network of ground receiving stations. Such satellites will not only record what they "see" and "feel" but will also receive and record data sent from the ground stations and, using on-board computers, perform the initial processing of the information gathered.

In 1968 and 1969 the Soviets launched at least five more weather satellites. These included Cosmos 184, 186, and 189 launched in 1968 followed by Meteora 1 and 2 launched in 1969. In general the use of weather satellites within the Soviet Union has reached operational status.

10 Lunar and Planetary Probes

The conquest of the solar system will not only give us energy and life, which will be two billion times greater than the earth's energy and life, but will give us spaciousness which will be even more abundant. We may say that man on earth commands two dimensions; the third is limited, i.e., propagation up and down is impossible at this time. When the solar system is conquered, man will have three dimensions.

K. E. Tsiolkovsky, 1929

THE LEAST successful of all the Soviet space launches have been lunar and planetary probes. Official Soviet sources have stated that twenty-two of their interplanetary probes with carrier rockets weighing 50 tons have been successfully launched into orbit during the first ten years of the cosmic era. United States' sources record, however, that the Soviets have launched at least thirty-six spacecraft to the moon, Mars, and Venus. Eighteen have been sent to the moon with approximately 50 percent success; seven have been probes to Mars with no successes; and the remaining twelve have been probes to Venus again with no successes in the first decade. The difference between the U.S. and Soviet totals are attributable to the Soviet policy of not announcing space failures, if at all possible.

135

Lunik

Most scientists agree that in order to understand the history of the earth-moon system, one must get firsthand knowledge of the properties and structure of the moon's surface. In particular, the investigation of the moon's inner structure, distribution of its internal mass, its thermal properties and magnetic field are essential to this understanding. Furthermore, they feel that there is a correlation between the moon and outer space and that many problems associated with the celestial bodies of the solar system can be solved by comparing the earth's geophysical and geochemical properties with those of the moon.

One of the first scientific objectives in the investigation of the moon was to determine the exact nature of the space just above its surface. This investigation was started with Luna 1, launched on January 2, 1959, fourteen months after Sputnik 1. Luna 1 was quickly christened *Lunik* by the American press, a name still in vogue today.

Although Luna 1 missed the moon by 3,100–3,700 miles and entered into a solar orbit with a period of fifteen months, it still enabled the Soviet scientists to establish that there was no perceptible magnetic field near the moon and to detect in planetary space streams of ionized plasma known as the solar wind. Both findings were important contributions to the fund of knowledge of interplanetary space.

Then on September 12, 1959, the Soviet Union launched Luna 2, which crashed on the visible side of the lunar surface some 270 miles from the center. Before the satellite crashed, its instruments sent back to earth the first direct information on the moon's physical properties and established that the earth's only natural satellite did not have a radiation belt.

Luna 3 was launched on October 4, 1959. Its main mission was to photograph the far side of the moon. It passed within 4,373 miles of the moon and photographed the eastern portion of the back side, stretching for 110° longitude. The time chosen for the experiment was when the sun was low above the lunar

FIG. 45   Luna 1, launched January 2, 1959; weight was 800 lb, diameter 47 in. Luna 2, launched September 12, 1959, was the same size, but weighed approximately 850 lb.

horizon. The height of the mountains in a photograph could then be judged by the shadows they cast, and the moon's other physical features could be seen in relief.

Since Luna 3 had covered only about 110° of the lunar longitude, about 70° longitude remained. Zond 3 was launched on July 18, 1965, and completed the photographic task. In addition, Zond 3 carried instruments for studying the magnetic properties of near-earth and interplanetary space, galactic low-frequency radio-wave emissions, the solar wind, cosmic rays, the infrarad and ultraviolet spectra of the lunar surface and for measuring micrometeorite impacts enroute.

Zond 3 traveled over the part of the moon that was turned toward the sun. The probe passed on its trajectory south of the lunar equatorial plane, which enabled it to cover more of the southern and greater portion of the unmapped part of the moon. The probe began taking pictures at an altitude of a little more than 7,000 miles.

137

FIG. 46 Luna 3 photographed the far side of the moon.

FIG. 47 Far side of the moon as relayed by Luna 3.

Zond 3 then passed over the far side of the moon. It gradually descended to the minimal altitude of 5,725 miles and then began gaining altitude again. Thirty-three hours after the blast-off, Zond 3 had photographed practically all of the areas unphotographed by Luna 3.

A new, compact photo and television system designed for operation during prolonged space flights was used in Zond 3. The system ensured protection of the film from cosmic radiation. The photo camera in the system had a focus of 105.4 mm and an aperture ratio of 1:8. Special 25-mm film was used in the camera. Exposure time was 0.01 sec and 0.03 sec. Film processing was done automatically while photographing was in progress, and processed film was fed straight to the television relay system. Television pictures were transmitted to earth via a directional parabolic antenna designed for long-distance communication.

The pictures provided a great number of details, especially regarding the structure of the lunar surface. They confirmed earlier conclusions that there were few so-called seas on the far side of the moon. While the north part of the visible hemisphere has numerous seas, the north part of the hidden hemisphere is a large continent specked with craters. This continent is much larger than the southern continent of the visible hemisphere.

Formations that are not found on the visible side of the moon have been discovered on its hidden side. These are chains of craters extending for hundreds of miles and sealike formations named *thalassoids*, vast, colorless depressions dotted with craters. In size, thalassoids are comparable to seas up to 310 miles in diameter. There are no gigantic depressions as clearly defined on the visible side of the moon. Over a thousand crater formations were detected on the far side. The newly discovered craters are typical of the moon as a whole. They fall into two types: craters from which streaks radiate in all directions, and craters with central peaks. The long crater chains are formed by medium-sized craters, 1.3 to 18.5 miles across. Some of these chains, which evidently radiate from the light-colored continent area north of the Eastern Sea, range over 372 miles and more.

On the lower right-hand side of lunar photographs there is a dark patch. A small portion of this patch is visible from the earth and has been known as the Eastern Sea, but its true size and shape became known only after the processing of information supplied by Zond 3. In close proximity to the Eastern Sea are the dark strips of Mare Automni and Mare Veris. Two more dark-colored formations were located there which cannot be seen from earth. Visual observations had led scientists to believe there was a sea, tentatively named the Shallow Sea, south of the Eastern Sea. This was confirmed by the Zond 3 data. The photographs furnished by Luna 3 and Zond 3 showed that some 40 percent of the moon's visible surface is covered

FIG. 48   Luna 9, with capsule shown lower left, the first rocket to make a soft landing on the moon, February 3, 1966.

FIG. 49   Pendant delivered to the moon by Luna 9.

by seas, while on the far side, seas account for only 10 percent
of the surface, are more elevated and lighter colored.

Newly discovered formations on the far side of the moon
have been named after outstanding men in Soviet and world
science. In 1960, the first atlas of the far side of the moon was
compiled on the basis of the Luna 3 photographs.

Between April, 1963, and December 3, 1965, the Soviet Union launched Luna 4 through Luna 8. All failed. Luna 4 missed
the moon by 5,281 miles and Luna 6 by 100,000 miles. Luna
5, Luna 7, and Luna 8 all had retroengine malfunctions, causing the probes to crash on the moon without accomplishing
their missions.

Finally, on January 31, 1966, the Soviets successfully soft-
landed Luna 9 on the moon. It descended in the Ocean of
Storms west of the Reiner and Marios craters. The probe had a
useful life of several days—communicating with earth by radio
and taking twenty-seven close-up pictures, including a panoramic view of the lunar surface.

The world's first artificial lunar satellite, Luna 10, was injected into orbit around the moon on February 3, 1966. The
orbital parameters were 610-mile apolune, 235-mile perilune,
and the inclination was 72°. It remained active in a lunar orbit
for fifty-six days (460 orbits) yielding a great deal of data.

141

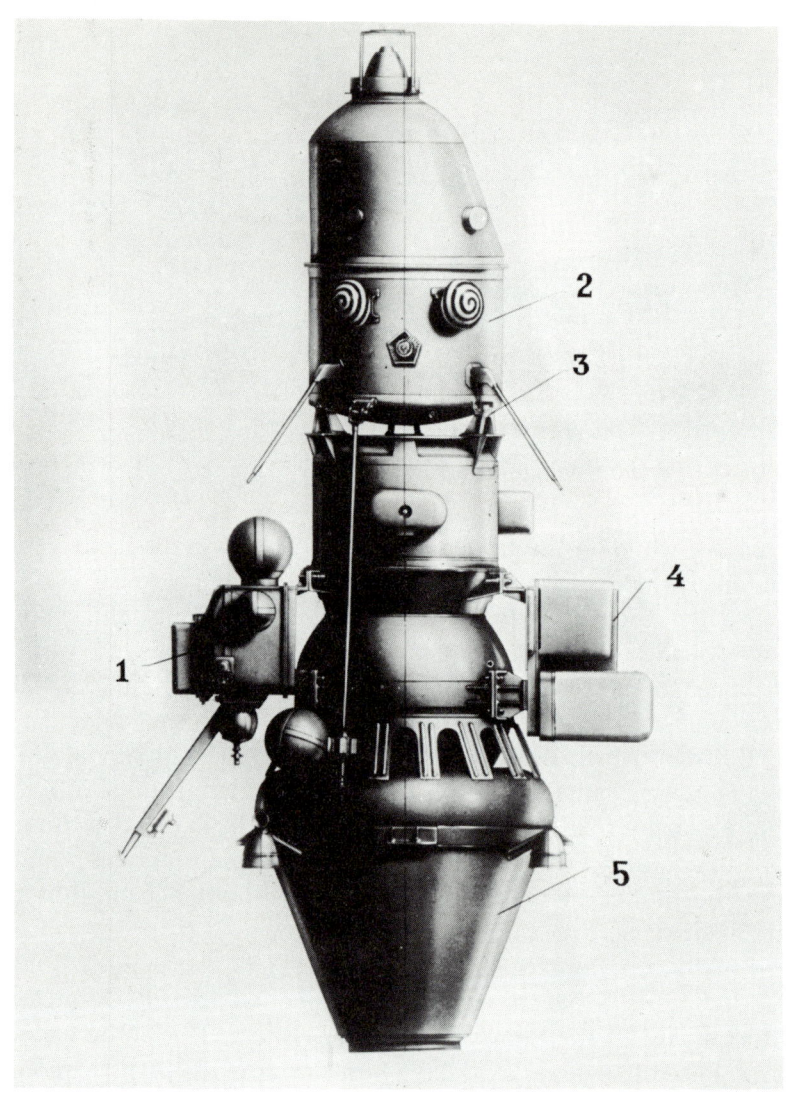

FIG. 50 Luna 10: 1) radio measuring system, 2) lunar satellite, 3) separation system, 4) astro-orientation system, 5) rocket engine.

The mission produced two significant achievements. First, the technology required for obtaining lunar orbits was developed to the point where other lunar orbital missions could be successfully accomplished. Second, it provided information on the γ radiation of the lunar surface, radiation conditions near the moon, solar plasma in cis-lunar space, infrared radiation of the lunar surface and data on the lunar gravitational field.

Prior to Luna 10's flight, it was thought the moon's shape was almost a sphere with a diameter of approximately 2,158 miles, with a slight elongation of the moon's body toward the earth. Measurements from Luna 10, however, indicated that the moon has a pear-shaped form with an elongation on the opposite side from earth.

Luna 10 determined for the first time the chemical composition of some lunar rocks by the nature of γ radiation of the moon's upper layers. Although the measurements were made over topographically different areas, the nature of γ radiation proved identical. The concentration of radioactive elements in lunar rocks was found to be close to that of our rocks.

Luna 10 showed that the moon's magnetic field changes from 24 to 38 γ whereas the earth's varies from 30,000 to 74,000 γ. These measurements were taken with the moon in different positions with respect to the earth and sun. During the full moon (as observed from the earth) the moon and Luna 10 were located in the area of the earth's magnetic "tail." No effects were recorded. Neither did the lunar magnetic intensity change with distance. From these observations, the Soviet experimenters concluded that the moon has no magnetic poles, and therefore the lunar magnetic field should be regarded as an interplanetary field deformed by the moon.

On August 22, 1966, Luna 11, the moon's second artificial satellite, was launched. Its orbital parameters were somewhat different from the first Lunik. It had a 102-mile perilune, 743-mile apolune, and an orbit close to equatorial as compared to Luna 10, which was in a near polar orbit. It made 277 orbits around the moon, and, in addition to investigating the moon and lunar space, Luna 11 recorded long-wave cosmic radio emissions.

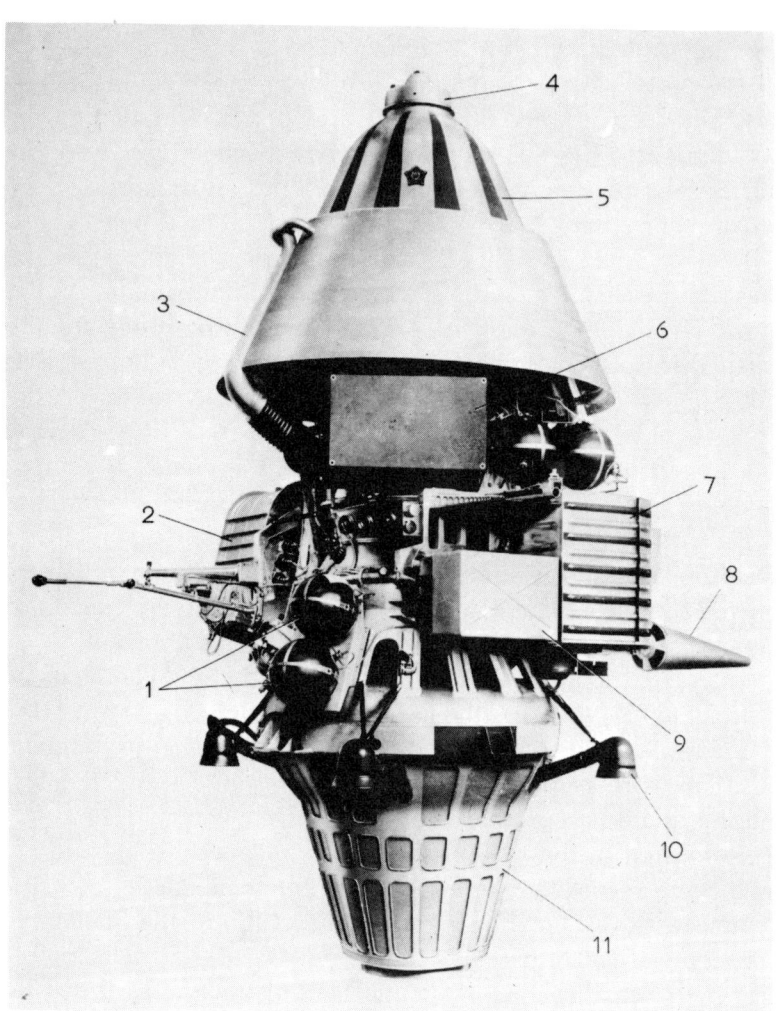

FIG. 51 Luna 12: 1) navigation pressurization system, 2) photo-television system, 3) thermal control radiator, 4) radiometer, 5) instrument compartment, 6) chemical battery, 7) optic-mechanical orientation system, 8) antenna, 9) orientation system electronics, 10) attitude control engines, 11) radio system measuring instruments.

FIG. 52 Pendant placed on board Luna 12.

Luna 12 went into lunar orbit on October 25, 1966. In addition to the investigations initiated by the first two Luniks, Luna 12 photographed sections of the lunar surface. The pictures were taken from a distance of 62 to 211 miles. Of special interest was the area near the Aristarchus Crater that, from earth photographs, appears as light rays emerging from the crater. The smallest objects that were identifiable from the Lunik photographs were about 12 miles wide. Soviet scientists concluded that the craters were so-called secondary craters, presumably formed by eruptions of fragments of rock from a volcano or by the fall of a meteor's body on the lunar surface.

Luna 13, launched in December 21, 1966, made a soft landing on the Sea of Storms. It carried a television camera which returned panoramic pictures of the lunar surface. An extendable test rod was driven into the lunar soil to determine its characteristics and bearing strength. A meter to record γ-radiation levels was also included.

Exploration of the moon continued into the second decade of the space age. After sixteen months of inactivity Luna 14 was launched on April 7, 1968, into a lunar trajectory and subsequently was injected into a 'unar orbit. Toward the end of

145

the year Zond 5 and Zond 6, as unmanned precursors for manned carrying craft, flew circumlunar missions and were successfully recovered after the flight.

The next year, 1969, proved to be one of the most active in Soviet space history as well as one of the most disappointing as far as their lunar program was concerned. Luna 15 launched on July 13, 1969, performed ingloriously just as the U.S. astronauts were landing on the moon. It subsequently failed on a lunar soft-landing attempt. Climaxing the year, the Soviets had two additional lunar failures identified only as Cosmos 300 and Cosmos 305.

With the 1968 Soviet Zond circumlunar flights, many people including NASA officials believed it possible for the Soviets to fly men around the moon before the United States and perhaps win the honor of making the first manned lunar landing. Apollo 11's landing on July 20, 1969, ended speculation for all time, as to whom the prize belongs for accomplishing the greatest and most far-reaching achievement of our time, that of landing man on the moon and returning him safely to earth. The actual moment of landing being immortalized by the words of Apollo 11's Commander Neil A. Armstrong as he swung down from the lunar module to the dusty Sea of Tranquility . . .

That's one small step for a man, one giant leap for mankind . . .

By the end of 1969, however, the big question was, "What had happened to the Russians?" Speculative answers focused on two possibilities. The first assumes that the Soviets have had considerable problems with their lunar launching vehicle. In fact, it is believed that the booster has been on the launching pad and actually proceeded into countdowns, but the count has never reached "zero." The second supports the premise that the super booster exploded on the launching pad killing many Soviet observers and scientists. Regardless what the reason, the Soviets lost the contest to be first.

It is now believed that the Soviets have laid aside their lunar aspirations to concentrate on earth-orbiting space stations. In fact, Mstislav Keldysh, President of the Soviet Academy of Sciences, has made recent public statements that can be interpreted that way. The recent Soyuz 6, 7, and 8 launches lends credence to these new Soviet objectives.

Mars Program

The Soviets' Mars program was a total failure. Flights began with unannounced launches in 1960, which failed to reach earth orbit. Two years later three more attempts were made to send probes to Mars. The first, again unannounced, exploded in earth

FIG. 53   Interplanetary station Mars 1 launched November 1, 1962.

orbit. The second, Mars 1, traveled 120,000 miles from the planet and then experienced the now familiar communications failure. The third failed to leave earth orbit.

During a 1964 planetary window opportunity, Zond 2 was launched. Although communications failed again, Zond continued on its trajectory and approached Mars within 930 miles. An interesting sidelight of this mission was the disclosure that Zond 2 carried an electric-jet plasma engine on board. This type of propulsion system appears to be promising as a control system for prolonged space flights.

Zond 3, which had as part of its mission photographing the back side of the moon, then continued on toward Mars in an engineering test program. This was the last of the Mars probes through the end of 1969.

Venus Program (*Venera*)

Probe failures continued to plague the Soviet scientists in the Venus program. The first two failures started with the launching of Sputnik 7 on February 4, 1961, and Sputnik 8 on February 12, 1961. These were "first generation" Venus probes weighing 14,300 lb in earth orbit.

Sputnik 7 achieved orbit but failed to launch its interplanetary probe. Eight days later Sputnik 8 successfully launched Venus 1 from its parking orbit, but fifteen days later, at a distance of 4,700,000 miles communication contact with the probe was lost. The probe continued its flight toward Venus eventually approaching within 12,000 miles of the planet. Skirting Venus, the probe became a satellite of the sun.

Since then, two unannounced failures occurred during the 1962 window and two during the 1964 window. The 1964 launches were identified as Cosmos 27 and Zond 1. When the 1965 launch window arrived, the Soviets were ready with three probes, Venus 2, Venus 3, and Cosmos 96. Venus 2 was launched on November 12, 1965, and Venus 3 on November 16, 1965. These probes, weighing 2,118 lb each, were put into identical heliocentric orbits and closely followed each other during the entire transplanetary flight. The advantage of a

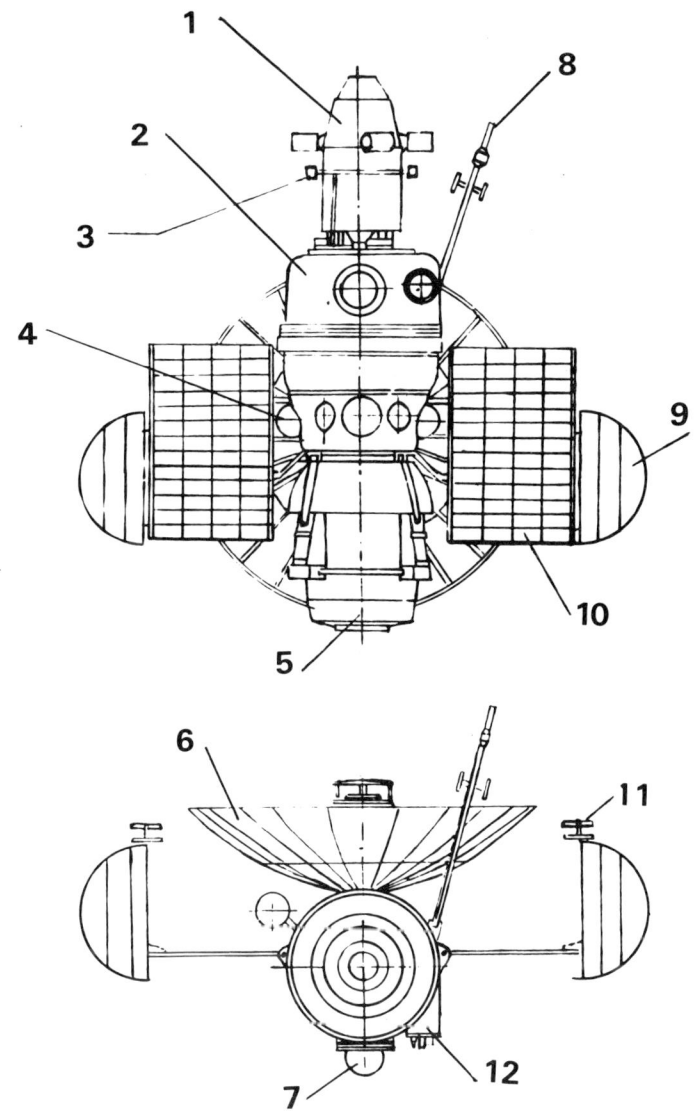

FIG. 54 Sketch of interplanetary probes Venus 2 and Venus 3: 1) Mid-course correction engines, 2) orbital compartment, 3) micro-engine orientation engine, 4) inert gas pressurization system, 5) special compartment, 6) parabolic antenna, 7) flight sensors, 8) magnetometer shaft, 9) thermal control radiators, 10) solar cell panels, 11) small directional antenna, 12) solar orientation sensor.

149

"team flight," as reasoned by Soviet scientists, was that it added to the reliability of the data obtained throughout the experimental flight. The data yielded by one probe was checked against the data recorded by the other. Cosmos 96 was launched on November 23, 1965, but failed to leave earth orbit.

Although the flight trajectories to Venus were identical, their missions at planetary approach were different and were phased at entry to yield the greatest amount of information. Venus 2's mission was a close flight near the planet surface. It was programed to measure certain physical planetary parameters and to photograph the planet's surface during its passage. Venus 3, on the other hand, was to enter the dense layers of the Venusian atmosphere and transmit back to earth temperature and near-surface pressure measurements.

On February 27, 1966, having traveled through space for 3.5 months, Venus 2 passed Venus at a distance of 14,964 miles. Because of a communications failure, however, no information was transmitted. A few days later, on March 1, 1966, Venus 3 crashed on Venus. But like Venus 2 it also experienced a communications failure, and no information was received. The Soviets claim that Venus 3 planted a sphere with the Soviet Union's national emblem on the surface of Venus.

On June 12 1967, the Soviets' only successful Venus probe, Venus 4 was launched. As with the earlier launches, the carrier rocket and probe were first put into an intermediate earth parking orbit. After orbital checkout (and at the correct time in the orbital path) the engines were activated, and the probe was put into a proper flight path toward Venus. The probe weight was 2,438 lb.

Trajectory corrections were required to assure Venus 4's landing on the Venus surface. Accordingly, a course correction was made on July 29, 1967, by the ground control when the probe was approximately 7.5 million miles from the earth. Telemetry information and trajectory measurements carried out after the flight course correction showed that the controls responded according to plan, and the spacecraft executed the proper maneuver putting it on the desired flight path to Venus.

Venus 4 was comprised of two principal parts: the orbital

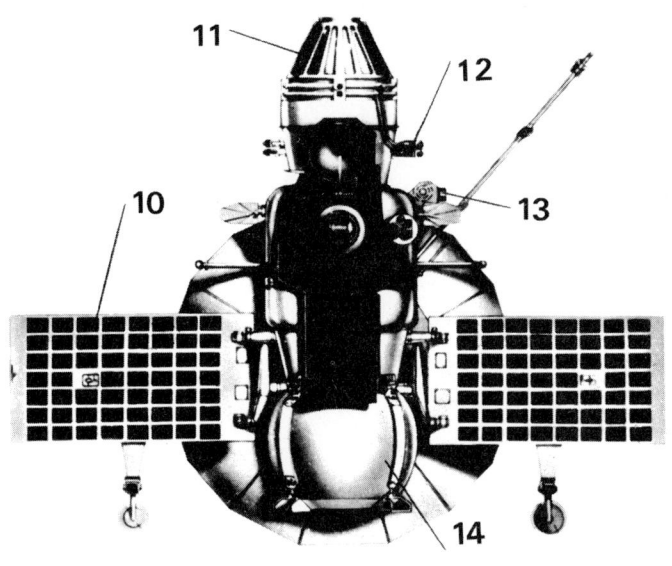

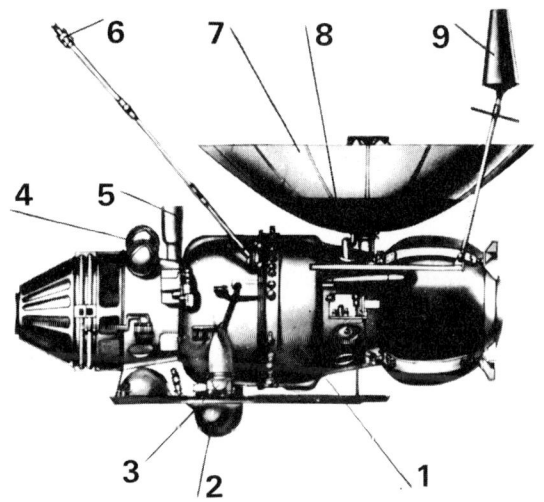

FIG. 55   Automatic interplanetary station Venus 4: 1) orbital module, 2) astro-orientation sensors, 3) solar orientation sensor, 4) pressurized inert gas spheres, 5) sun-earth orientation sensor, 6) magnetometer, 7) parabolic antenna, 8) thermal control radiator, 9) low directional antenna, 10) panel of solar cells, 11) mid-course correction engine, 12) orientation micro engines, 13) cosmic ray particle counter, 14) Venus reentry capsule.

FIG. 56 Pendant delivered to surface of Venus by Venus 3.

compartment, and the soft-landing capsule. The orbital compartment contained a midcourse correction engine, instrumentation sensors, antennae, electrooptical sensors for the altitude-control system, solar batteries, and vernier motors. The pressurized container of the orbital compartment accommodated electronic equipment of the probe's various systems, power supply, and temperature control units. The temperature-control system maintained ambient temperature of between 40.3° and 47.4°F.

A month before the probe reached Venus, the place and time of entry into the Venusian atmosphere were determined, and a program for communications was sent to the probe by the ground control station. Communications began on October 18, 1967, approximately 117 min before the soft-landing capsule was released some 28,000 miles from Venus.

During the communication periods the probe used a directional parabolic antenna, which necessitated high pointing accuracy toward earth. Accurate attitude control had to be maintained during the trajectory adjustments. Orientation of the probe was achieved by the altitude-control system, which was

152

comprised of optical sensors, gyroscopic instruments, and electronic controls. Tiny vernier jet motors serve to maneuver the
probe in the required direction. The probe's attitude in space
was determined in relation to the earth, the sun, and the star,
Canopus. The solar-cell panels were oriented perpendicular to
the sun rays most of the time during the flight. A special sensor
served to maintain that sun-oriented attitude of the probe.

On October 18, 1967, the soft-landing capsule was separated
from the probe. The entry capsule was practically spherical
in shape, with a diameter of about 40 in. and weight of 843 lb.
Within this spherical configuration were two pressurized compartments, one for the instrumentation and the other for the
descent parachute.

The capsule on entering the Venusian atmosphere experienced atmospheric braking and a deceleration of about 300 g.
To protect the capsule from aerodynamic heating, its surface
was coated with a special ablative coating. As the capsule

FIG. 57   Venus 4 reentry capsule.

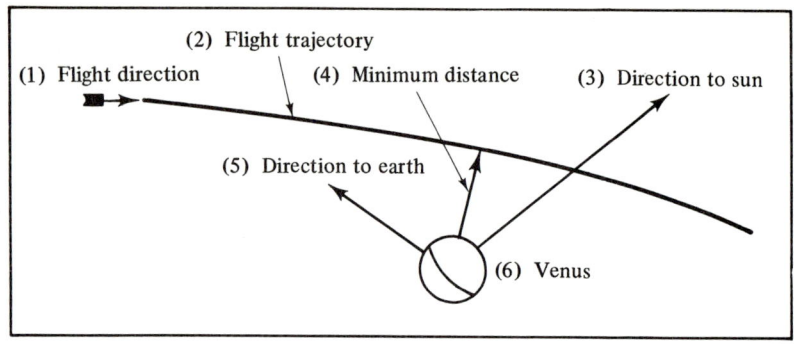

FIG. 58 Flight trajectory of Venus 2 and Venus 3. Venus 2 passed planet at a distance of 14,500 miles; Venus 3 impacted on Venus.

slowed down to about 670 mph, parachutes were automatically opened by a system comprised of sensors to monitor atmospheric pressure and *g* forces and a timing mechanism. When the main parachute opened, the antenna system, the radio altimeter, and the telemetry transmitter were activated, and the capsule began to transmit Venusian atmospheric data to the receiving station on earth. The capsule was then about 16 miles above the surface.

During the initial phase of the parachute's descent, the speed was about 33 ft per sec. As the descent progressed and the atmospheric density increased, the speed of the capsule decreased to 10 ft per sec, which would permit a soft landing. The probe entered the atmosphere on the planet's nocturnal side.

According to Soviet reports pressure, density, temperature, and the chemical composition of the Venusian atmosphere were constantly monitored during the descent. The Soviet readings showed that the atmospheric temperature ranged from 104°F to 536°F and the pressure from 1 to about 15 atmospheres. They also showed that the Venusian atmosphere

FIG. 59 *Facing page:* Venus 4 capsule flight and descent into Venus atmosphere: 1) approaching Venus radio contact, 2) separation of reentry capsule, 3) deceleration of reentry capsule into atmosphere, 4) opening of braking parachute, 5) opening of main parachute, 6) beginning of radio transmission, 7) capsule on surface after landing.

154

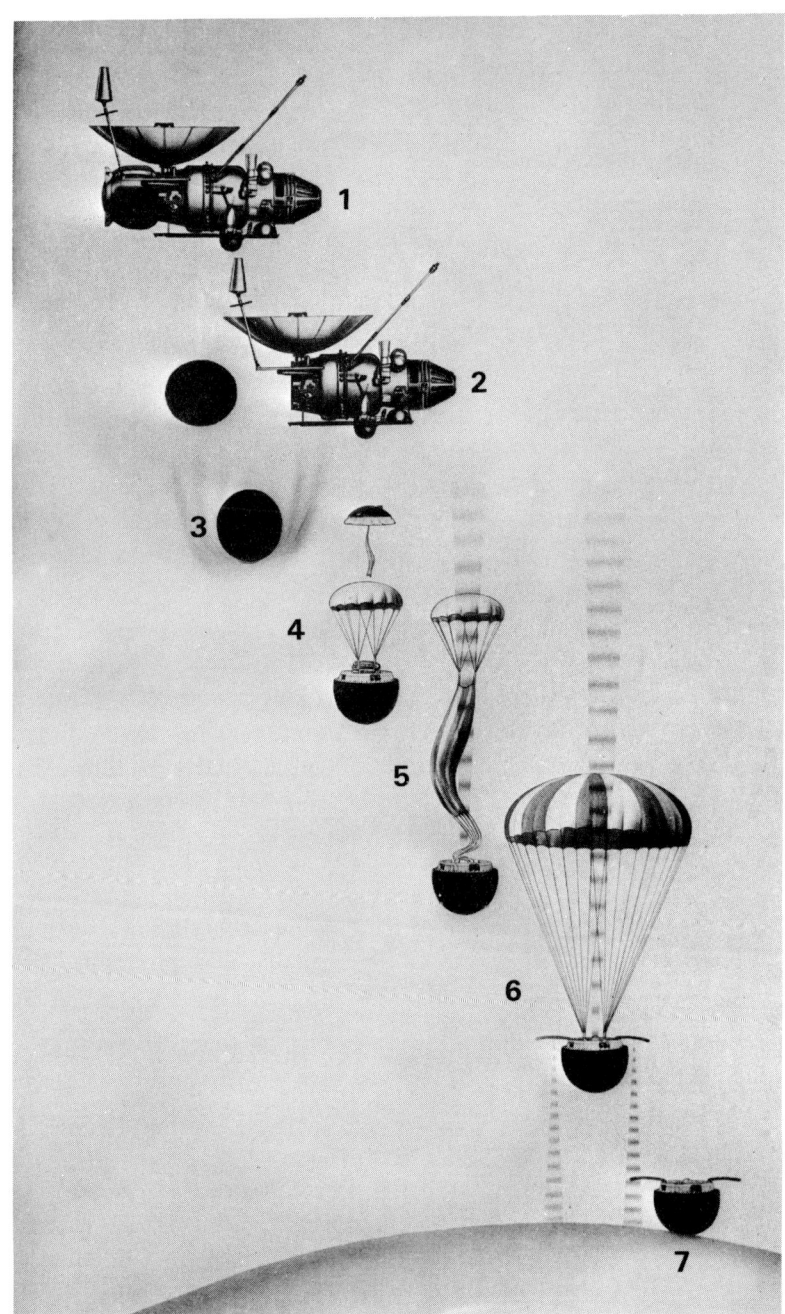

consisted almost entirely of carbon dioxide; oxygen and water vapors constituted about 1.5 percent; and no trace of nitrogen was found. The instrumentation aboard the capsule measured the atmospheric parameters continuously for about 90 min while descending the 15 miles. At 09 hr 14 min (Moscow time) the telemetry signals stopped. The capsule delivered a second sphere on the surface of Venus, also containing the Soviet Union's emblem.

The Soviet exploration of Venus continued with the dual launches of Venus 5 and 6 in January, 1969. After a 130-day journey the spacecraft entered the Venusian atmosphere on May 16, 1969, and May 17, 1969, respectively. The scientific capsules made a smooth descent into the atmosphere for a period of 51 minutes.

Commenting on the flight of the spacecraft and how Venus 5 and 6 differed from Venus 4, the new Chief Designer said,

In many ways they are of one type; however, there are significant differences. For example, Venus 5 and 6 had more refined electronic instrumentation and a new radio-altimeter. The main differences, however, were in the descent package. The latter were designed with a consideration of the Venus 4 results. In particular, the descent packages for the new stations were designed to withstand far greater g [gravity] loads than their predecessor; they can withstand g loads of 450 units. . . .

When we launched Venus 4 we knew very little about what to expect of its atmosphere. After the flight a more exact model of the atmosphere was established. Venus 5 and 6 also carried different parachutes . . . three times smaller than on Venus 4. As a result the rate of descent increased.

Commenting on the preliminary results received from the spacecraft, the Chief Designer continued with:

Now, as a result of the investigation, it is clear that this planet is hardly suitable for man. In other words, our children and grandchildren will not fly there. But for an understanding of the planet's structure and evolution, such experiments have great value. This [flight] was also necessary for the development of space technology and for further flights in space. After the Venus flights, for example, we can say that automatic spacecraft can "operate" in regions even closer to the sun; e.g., to such planets as Mercury.

An interesting epilogue to the Soviets' exploration of Venus by Venus probes had its beginning about thirteen years ago when Cornell H. Mayer, a radio astronomer at the Naval Research Laboratory in Washington, D.C., first directed his fifty-foot diameter radio telescope dish at Venus. Mayer recorded strong emissions from Venus from which he estimated the surface temperature to be about 620°F. These findings were questioned both here and abroad.

In 1962, when the U.S. Mariner 2 flew by Venus, its instruments confirmed a high surface temperature. From this and other data, U.S. scientists constructed a planetary atmospheric model which also indicated a high atmosphere pressure.

Unconvinced—some Soviet scientists still believing that the high temperature readings came from Venus' ionosphere—their engineers were instructed to build the Venus entry capsules to withstand only relatively low pressures. As a result, the three Soviet probes that eventually were launched to Venus (Venus 4 in 1967 and Venus 5 and 6 in 1969) crushed like egg shells as they descended into the atmosphere on parachutes below about fifteen miles. During their descent, however, they radioed back important information confirming that carbon dioxide (CO_2) makes up about 95 percent of the planet's atmosphere and that a high planetary temperature exists.

The Soviets later admitted the error. Scientists extrapolating both U.S. and Soviet data now agree that Venus' surface pressure is about one hundred atmospheres and the surface temperature about 900°F and not the 20 atmospheres pressure and 530°F temperature as originally reported. The earth's pressure as sea level is one atmosphere.

11 The Cosmonauts

We shall need many fresh and en-
thusiastic teams. Flights to the
stars cannot be compared with
flights in the air. . . . If they un-
derstood the difficulties, many now
working with enthusiasm would
abandon it.

K. E. Tsiolkovsky, 1929

THE TRANSITION from unmanned orbital flight to manned
orbital flight began after the successful launching of the Soviet
Union's third unmanned satellite on May 5, 1958. This flight
precipitated the need for training a group of spacemen. Lt.
Gen. Nikolai Petrovich Kamanin was selected to lead this new
branch of Soviet aviation as Deputy Chief of the Soviet Air
Force.

Nikolai Petrovich Kamanin

General Kamanin was born on October 18, 1908. His education included the Air Force Theory school in Leningrad
(1927–1928), a pilot training school in Borisoglebsh (1928–
1929), the Zhukovsky Air Force Engineering Academy (1934–
1938), and the Academy of the Soviet Armed Forces General
Staff (1956).

In 1929, Kamanin became the first Hero of the Soviet Union.
That award, the nation's highest, went to him for saving the
crew and passengers of the ice-breaker *Chukyuskin*. Several
months after his successful rescue operation, he entered the
Zhukovsky Academy and shortly after graduation became a
Soviet Air Force Commander. For several years Kamanin was
in command of an air brigade in the Ukraine.

From 1942 to the end of World War II, he was commander of an attack aircraft division at the Kalinin Front. Later, as commander of an attack aircraft corps, Kamanin participated in routing Nazi troops at the Kursk defenses near Kharkov, Poltava, Kiev, Lvov, at Sandimir in Poland, as well as in Rumania, Hungary, Czechoslovakia, and Austria.

For the first two postwar years, he served as Deputy Chief of the Aeroflot (Soviet Civil lines) and from 1947 to 1954, he was Chairman of the Presidium of the Voluntary Society for Promotion of Aviation (DOSAV), working toward the revival of aero-clubs and the development of aerobatics gliding, parachute jumping, etc., in the Soviet Union. Many Soviet astronauts were students of aero-clubs organized by Kamanin. His success at his present assignment is best reflected in the many "firsts" of the Soviet manned space program and also by the fact that each of the cosmonauts under his command has been made a Hero of the Soviet Union.

Yuri Alexeyevich Gagarin

On April 12, 1961, Cosmonaut Yuri Gagarin, aged twenty-seven, became the first man to orbit the earth. During his brief 108 min in orbit, Gagarin confirmed that man can exist in space in a weightless condition and perform assigned tasks.

Gagarin's first step toward outer space, he said, was joining an aero-club. His second step was attending the Orenburg Pilot Training School, which intensified his interest in flying and in outer space.

After the successful launching of the Soviet automatic station, Luna 3, Gagarin, by then a fighter pilot in an air force unit, applied for admission to the astronaut training group, "if such a group exists." Some time later, when the group's instructor asked the prospective cosmonauts whom they thought should be the first man to orbit the earth, about 60 percent named Yuri Gagarin.

After his successful orbit Gagarin was given command of the astronauts' group and was elected an honorary member of the International Academy of Astronautics. He gave up

FIG. 60 S. P. Korolyov and Cosmonaut Y. A. Gagarin.

command of the cosmonauts in 1964 and attended the Zhukovsky Air Force Engineering Academy until 1967.

On March 27, 1968, Yuri Gagarin died in an airplane crash along with Col. Vladimir S. Seryogin, a test pilot, some forty miles northeast of Moscow.

Gherman Stepanovich Titov

In April 1961, Col. Gherman Titov was the cosmonaut understudy of Yuri Gagarin, and four months later on August 6, 1961, he became the second man to orbit the earth. Titov was born on September 11, 1935, is married, and has two daughters. He attended pilot training school at Volgograd and the Zhukovsky Air Force Engineering Academy. His astronaut career followed service in the Air Force, with many speed and high-altitude flights to his credit. He joined the astronaut group in 1959 and was second on the list of men suggested by the cosmonaut team members to pioneer the first manned space flight.

While Gagarin led the way into outer space, Titov proved man not only can exist in the cabin of a spaceship, but also can work very effectively. He was the first man to take motion pictures, pilot his spacecraft by manual control, eat and sleep

160

normally, undergo different biological and psychological tests and do physical exercises while in orbit.

Titov's 24-hour flight around the earth went well. During the sixth and seventh orbits he suffered symptoms resembling seasickness, as every movement of his head caused light-headedness and nausea. Because of this, Soviet scientists made some changes in the system of training astronauts. A special flying laboratory where the condition of weightlessness was maintained for from 20 to 25 sec was introduced. In subsequent Soviet manned space flights, the volume of medical research was enlarged and investigation of the ear (or vestibular) apparatus expanded.

Titov is Deputy Editor-in-Chief of the *Aviatsio i Kosmonavtiko* (Aviation and Astronautics), the Soviet Air Force monthly and has been commander of the cosmonaut group since 1964. He is credited by the Soviets as saying: "I have lived seventeen days longer than any of you. I have seen the sun go up and down seventeen times in one day."

FIG. 61   Cosmonaut Titov being tested for reactions in centrifugal chamber.

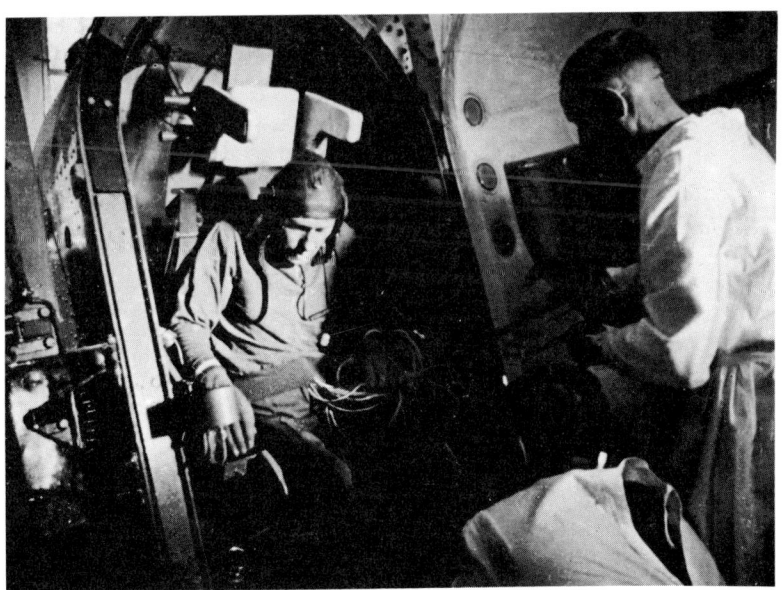

Andrian Grigorievich Nikolayev

Colonel Nikolayev, Titov's understudy, flew Vostok 3 from August 11 to August 14, 1962, becoming the first cosmonaut to make a four-day orbital flight around the earth. During three of those days, Pavel Popovich flew Vostok 4 in close proximity to Vostok 3.

According to the Soviets, there were several reasons why the third manned orbital flight had to be a dual flight. The experience of Titov had made it clear that man's reactions to the flight conditions are diverse and complex. For the future of astronautics it was necessary to find out whether or not Titov's nausea was the inevitable result from protracted weightlessness or due to individual physical peculiarities of the astronaut.

Compared to the scope of assignments set to Gagarin and Titov, the program of the first dual Soviet flight was extended to include many complicated and sophisticated tasks. In the

FIG. 62   Cosmonauts Nikolayev and Popovich.

FIG. 63　Popovich and Nikolayev during training.

nearly ninety-five hours of spaceflight in the Vostok 3, Nikolayev covered a distance of over 1,240,000 miles, performing
the entire mission assigned to him and providing science with
the needed data as to what man is capable of during a lengthy
space flight. Experts in astronautics became convinced that
most technical problems associated with orbital rendezvous in
near-earth orbit could be solved.

Nikolayev was born on September 5, 1929. He attended a
fighter pilot training school and the Zhukovsky Air Force Engineering Academy. He joined the cosmonaut group in 1959
and was assigned "standby" pilot for Titov prior to his own
flight in Vostok 3. In November, 1963, he married Valentina
Tereshkova, the world's first woman cosmonaut.

Pavel Romanovich Popovich

Colonel Popovich flew Vostok 4 on the dual flight with Nikolayev. He stayed in space for 71 hours, orbiting the earth over
forty-eight times and covering a distance of approximately
900,000 miles. He participated in the vast research program
with Nikolayev.

Pavel was born on October 5, 1930. He attended a pilot
training school and the Zhukovsky Air Force Engineering

Academy and after an eight-year service in the Air Force, joined the astronaut group in 1959.

Valery Fyodorovich Bykovsky

Aboard the Vostok 5, Colonel Bykovsky was the first to make a five-day orbital flight, from June 14 to June 19, 1963, around the earth. For three of those five days he flew a dual mission with Vostok 6, piloted by Valentina Tereshkova. Bykovsky made more than eighty-one orbits around the earth covering a distance of over 2 million miles.

During his 121.5 hours in space, Bykovsky conducted scientific research on himself to make certain what conditions were needed for man's short and prolonged stay in space. With that aim in view, Bykovsky navigated his ship; controlled the work of on-board systems; periodically established radio communication with earth and Vostok 6; made observations of the sun, the moon, and the stars; regularly made physiological, vestibular, and psychological tests; did special physical exercises; and conducted complex scientific investigations that required a high degree of precision. Specifically, he did three fixations of the *Tradescantia paludosa* microspores to find out the effect of dynamic stress, weightlessness, and radiation on the hereditary structures of this biological specimen.

Bykovsky's flight vastly enriched both space biology and space medicine. Using the results of research conducted by Bykovsky, as well as the similar research done by Valentina Tereshkova, scientists traced a series of concrete changes in the functional condition of the blood circulation system as related to the action of weightlessness and grouped together these changes as phases depending on the length of flight. Soviet scientists wished to find out to what extent prolonged weightlessness changed the normal condition of the heart. Knowing this, it was possible to estimate heart deviations during the descent. This was the purpose of a later experiment with two dogs on a twenty-day orbit of Cosmos 110.

Bykovsky was born on August 2, 1934. Prior to joining the astronaut group in 1959, he served as a second-class military

pilot with the Air Force. In August, 1962, he was the standby
pilot for Nikolayev, captain of Vostok 3. Later, he attended
the Zhukovsky Air Force Engineering Academy.

Valentina Vladimirovna Tereshkova

The world's first and only woman cosmonaut to date, Valentina Tereshkova, made a three-day dual flight aboard Vostok
6 from June 16 to June 18, 1963. The companion spacecraft was
Vostok 5, piloted by Valery Bykovsky.

During her 71 hours in the spacecraft, she participated in
an extensive flight program, as well as making physiological,
vestibular, and psychological tests, doing special exercises in
a weightless condition, and maintaining radio communication
with the earth and Vostok 5.

Tereshkova's flight provided science with data on reactions
of a female organism in space flight and confirmed the Soviet
scientists' assumption that women can explore space as well
as men.

Born on March 6, 1937, Tereshkova is married to the cosmonaut Andrian Nikolayev. Like the other cosmonauts, she at-

FIG. 64   Cosmonaut Valentina Tereshkova.

tended the Shukovksy Air Force Engineering Academy. She joined the astronauts' group in the spring of 1962, after classes at the Yaroslave Air Club, in which she won first place in parachute jumping.

Vladimir Mikhailovich Komarov

Komarov has the distinction of being the commander of the world's first multiseater spacecraft, Voskhod 1, which made a 24-hour flight around the earth on October 12, 1964. The crew included a research worker, Konstantin Feoktistov, and a medical doctor, Boris Yegorov. This was the first space crew consisting of people of different professions, and it was hoped they could meaningfully increase the amount and scope of research on long flights, particularly in reference to trips to other planets. After sixteen orbits around the earth, Komarov said the multiseated Voskhod made it possible to conduct complex physicotechnical and medicobiological research needed for interplanetary travel and that having a crew aboard the spacecraft makes it possible to obtain more objective data.

Komarov was born on March 16, 1927, and died on April 23, 1967, as a result of an accident during the reentry of Soyuz 1 spacecraft. He is buried in Red Square, Moscow.

He attended a specialized secondary Air Force school in Moscow; a primary pilot training school at Sasovo; the Chkalov Air Force school at Borisoglebsh; the Zhukovsky Air Force Engineering Academy, where he also took a postgraduate course. He joined the astronauts' group in 1959, after serving in the Air Force. In August, 1962, he was the standby pilot for Popovich. In both 1964 and in 1967, Komarov was made a Hero of the Soviet Union, and he received many Soviet and foreign decorations.

Cosmonaut Gherman Titov paid this tribute to Komarov after his death:

> The system that had heretofore operated perfectly failed to open [i.e., the system that deployed the parachute]. The price for this misfire was excessive.
>
> Komarov was the most experienced and prepared among us;

he was twice given tests of new spaceships: first the Voskhod, and then Soyuz 1. In his last flight, he completely fulfilled the sophisticated test program and the scientific experiments. Right up to the final minute, he reported the operation of the numerous systems and mechanisms, transmitting the readings of the instruments.

Somewhere far above Africa, on the night side of the planet, all the necessary preparations for landing were made. The ship was oriented and the retroengine unit was switched on. He reported to the control point how many seconds this unit was in operation and how the separation of the instrument package from the ship occurred.

Komarov's voice was heard before the ship's entry into the dense layers of the earth's atmosphere. His report is a model of intelligent, succinct information and self-control. The ship descended precisely in the region intended for landing and was immediately detected by search planes and helicopters. Its descent velocity, however, seemed to be greater than tolerable. . . .

Konstantin Petrovich Feoktistov

Born on February 7, 1926, Feoktistov is the first space scientist to have participated in an earth-orbital flight. As a member of the crew of the world's first multiseater spacecraft, Voskhod 1, he made the 24-hour orbital flight around the earth with Komarov in 1964.

In the flight, his strictly defined functions complemented those of Komarov and the medical doctor, Boris Yegorov. As planned, Feoktistov made extensive observations of the sky. He found out that it was possible and convenient to orient the craft by the stars and that astronavigational measurements could be made with the help of a sextant. He helped future interplanetary navigators by concluding that a spacecraft's position in the universe can be determined and its flight trajectory calculated right on board the spacecraft. Belayev and Leonov, who flew Voskhod 2, say that the results of Feoktistov's experiments helped them prepare for Leonov's first walk in space.

Feoktistov developed a passion for space back in his boy-

FIG. 65   Cosmonaut Yegorov in attitude chamber.

hood. At the age of ten, he worked out a detailed plan to explore the moon, which involved studying mathematics and physics in order to understand the Tsiolkovsky formulas. He faithfully followed his plan through school and college and became well-versed in Tsiolkovsky's theories. In 1955, Feoktistov completed his M.S. degree in engineering. After his space flight, he wrote a thesis that earned him a D.S. degree.

Boris Borisovich Yegorov

Yegorov, the first medical doctor in space, participated in the three-man flight aboard the spacecraft Voskhod 1 with Komarov and Feoktistov. As the ship's doctor, Yegorov had a busy program to fulfill. Well-equipped to conduct investigations and make measurements, he studied the condition of the cardiovascular system and watched the reactions of the central and peripheral nervous systems of the crew members.

Yegorov recorded brain biocurrents and electric potentials that emerged during spontaneous and deliberate movements of the eyes, calculated the parameters that characterize motor coordination in the process of drawing and writing, and collected data on muscular capacity for work during rhythmic movements of hands. In addition to the medical and biological investigations, Yegorov performed routine duties aboard the spacecraft.

Yegorov made the flight equipped with impressive knowledge of space medicine. He had investigated sources of irritation of the vestibular apparatus, suggested subtle micromethods of studying individual cells of the vestibular apparatus, and conducted many tests on animals, as well as investigating the causes of motor disturbances in a laboratory of a research institute of the U.S.S.R. Academy of Sciences and in Air Force units. As a sportsman, he flew planes and made parachute jumps.

Yegorov was born on November 26, 1937, in Moscow. He attended the First Moscow Medical College and earned his M.S. degree in medicine in 1967.

Pavel Ivanovich Belayev

Belayev was commander of the two-man version of Voskhod 2, which on March 18, 1964, made a 24-hour orbital flight around the earth. During the flight, the second pilot, Alexei Leonov took the world's first space walk.

During the flight of Voskhod 2, Belayev, like his Soviet predecessors in space, conducted astronavigational investigations and experiments. He confirmed the observations made by the Voskhod 1 crew under Komarov's command as to the presence of different layers of brightness in the atmosphere. The Belayev-Leonov team carried on the research begun by Dr. Yegorov to establish the sensitivity threshold of the vestibular apparatus and reactions of the nervous system in a weightless condition.

In addition to supervising Leonov's walk in space, Belayev was allowed to make the first touchdown by manual control. Heretofore, during the flight of the six spacecrafts of the Vostok series and of the first Voskhod, the control of the spacecraft during reentry was handled by computer.

Belayev was born on June 26, 1925. He attended a school for naval fighter pilots in Yeish, the Air Force Academy in Moscow, and took a postgraduate course at the Academy. During the same year, 1959, Belayev joined the astronauts' group and was its first commander until the summer of 1961, when he broke his leg in a parachute jump and required a year's recovery.

Alexei Arkhipovich Leonov

Leonov was the second pilot of the two-man spacecraft Voskhod 2 and the first man in history ever to walk in outer space. Asked by Soviet journalists before the launching why Leonov was to "walk" outside the spacecraft, the Chief Designer Academician Sergei Korolyov, explained:

> The walk-out is required for the many observations that are easier to make in open space than from aboard a spacecraft. The craft may need repair in flight. . . . A need may arise to change

FIG. 66   Cosmonauts Belayev and Leonov.

FIG. 67 Cosmonaut A. A. Leonov in flight suit.

crews. I visualize that then the new crew will be sent out in space aboard a simplified vehicle and the change of crew will be effected through lock-chambers . . . the coming experiment is another step in this direction.

Other Soviet experts believe that the Leonov experiment was a step toward the future when people in space suits will assemble interplanetary stations, space observatories, and large interplanetary ships with a view to reaching remote planets of the solar system and other galaxies.

In his ten-minute walk in space, Leonov covered almost four thousand miles of Soviet territory. After returning to the safety of the spacecraft, he worked at duties identical to those of his commander, Belayev.

Leonov was born on May 30, 1934. He attended a pilot training school, entered an air force school in Chugugev, and was subsequently admitted to the Zhukovsky Air Force Engineering Academy.

Leonov joined the astronauts' group in 1959 and has been editor of its newspaper ever since. In 1965, he was appointed the group's second in command. Prior to his space career, he served for three years in the Air Force as a Third Class Military Pilot.

12 The Flights and the Walk

From the rocket we can see the huge sphere of the planet in one or another phase like the moon. We can see how the sphere rotates and how within a few hours it shows all its sides successively. . . . This picture is so majestic, attractive and infinitely varied that I wish with all my soul that you and I could see it. Every two hours the rocket is eclipsed and enters the earth's shadow and night. This latter lasts for less than an hour; then for over an hour the sun shines before giving way again to darkness.

K. E. Tsiolkovsky, 1911

"HOW BEAUTIFUL is our earth!" These were the words of Cosmonaut Yuri Gagarin to describe his feelings on April 12, 1961, when he was the first man to see our planet from outer space.

Before Gagarin was sent into orbit, Soviet scientists made certain that the flight would neither be harmful nor cause basic changes in man's physical condition. Accordingly, some of the more critical biological problems were investigated prior to the flight by conducting experimental flights with animals first in high altitude rockets and later in satellite spacecraft. The main source of anxiety was the question of how a living organism would stand up to the high launch accelerations, how it would react to weightlessness, and how it would adjust

172

from the weightless condition to the high accelerations that occur during the orbital reentry phase. The animal experiments confirmed their suppositions that the living organism could withstand these and other factors of space flight without hazard to health.

It was only after this had been made certain that Gagarin was sent into outer space. In the final preparations, Sputniks 4, 5, 6, 9, and 10 were first launched. The primary purpose of these flights was to make a thorough check of the on-board systems, so as to ensure safe and controlled orbital flight and return to earth. During these flights, animals were tested for their physical reaction to space, and important investigations in the physics of outer space were also conducted. The information, particularly measurements of absorbed doses of radiation inside the spacecraft, was indispensable for the ultimate step, man's entry into space.

Sputnik's design also provided scientists with an opportunity to carry out medicobiological investigations during all phases of space flight, as well as the preflight and postflight examinations of test animals, thus studying the over-all physical consequences. The flights of Sputnik 5, 6, 9, and 10 were actually experiments of this kind. Investigations were carried out with a large number of living organisms. A number of biological objects of increasing complexity were placed in the spaceship capsule, beginning with microorganisms and plants and winding up with dogs. On these grounds, Sputnik 5 and Sputnik 6 were termed "flying laboratories." The experiments with Sputnik 9 and Sputnik 10 were a dress rehearsal for manned space flight. These test spacecraft were put into similar orbits selected for the actual flights that were to be flown by the cosmonauts, and the duration was also the same as planned for the first manned flight.

Vostok Spacecraft

Man first went into space in a Vostok spacecraft designed by the Korolyov team. It was a one-man ship, consisting of a spherical capsule, or command module, and instrument com-

FIG. 68   Vostok 1 spacecraft.

partment joined together by four bracing bands. Its weight with the last stage of the carrier rocket was 6.2 tons (4.7 tons without the last stage), and its length with the last stage was 24 ft. The command module had a diameter of 10.5 ft and a weight of 2.4 tons, and its outer surface was covered with a special heat-proof coating that protected it from the effects of high temperature when it entered the dense layers of the atmosphere during the entry phase.

The command-module shell had three hatches: the parachute hatch, the technological hatch, and a circular exit hatch. It also had three portholes of refractory glass. Through these portholes the cosmonaut could observe outer space, the earth, and take pictures. To protect the cosmonaut's eyes from sun rays, each porthole had shutters that could be closed

174

if needed. The porthole directly in front of the cosmonaut had a Vzor optical device affixed to it to assist the cosmonaut in manually aligning the spacecraft.

Located in the front of the cabin, over the optical device, was the instrument panel, and somewhat below it a television camera giving an inside view of the cabin. The instrument panel included gauges that measured and displayed the temperature, pressure, and content of the air in the cabin, the gas pressure in the orientation-system tanks, and other parameters. The panels also included a clock, which was switched in at launch and operated throughout the entire flight to the moment of landing.

The upper section of the panel was equipped with a powered globe of the earth rotating on two axes, which was used by the cosmonauts to determine visually the spacecraft's location in orbit and to determine the geographical coordinates of the assigned landing areas. The globe rotated about the two axes simultaneously. One of the axes corresponded to the earth's rotation; the other to the movement of the ship in orbit.

Located to the left of the cosmonaut were the container with the landing-system parachutes, a supply of drinking water, the ship's control panel, the emergency heat-regulation system, the landing-system direction finder, and a tape recorder. The right-hand side of the cabin contained the pilot's control stick, the food container, air-regeneration system, electric clock, cabin-side-view television camera, radio receiver set, sanitary system, and the battery electric-power-supply units capable of providing power to all the spacecraft equipment for a flight duration of ten days. All told, the Vostok had about 240 electric bulbs, 56 electric motors of various sizes and functions, over 6,000 transistors, and about 800 relays and switches.

The cabin's environmental-control system maintained the atmospheric gas mixture, consisting of 24 percent oxygen, plus other gases such as nitrogen, helium, and carbon dioxide in the proper proportions, at the correct pressure of 15 lb per sq. in., or slightly above sea level, and at the proper relative humidity, between 30 to 70 percent, all at comfortable gas-air temperatures. The air-regeneration system was carried out

FIG. 69 Vostok 1
spacecraft details.

automatically with the help of special chemical compounds.

The heat-regulating system also worked automatically. It consisted of an air and a liquid circuit connected to a heat radiator installed in the instrument module. Both circuits converged in the heat-exchange unit in the ship's cabin. The cosmonaut made the initial temperature adjustment in the spacecraft (from 54° to 77°F), which was then maintained automatically. This temperature setting was maintained even during the prolonged heating of the capsule's surface during the entry into the atmosphere.

The orientation system was used for positioning the ship in space when in orbit and during descent. The system could be operated both automatically and manually. It included actuating units, sensitive gyroscopic and optical elements, and logic devices. The actuating units in the ship's orientation system were the microjet rocket engines installed on the outer surface of the spacecraft.

The radio communication equipment included a signal system operating on a frequency of 19.995 mc. It served to determine the ship's position and to transmit part of the on-board data. The communication link also included a two-way radio telephone (two short-wave channels 9.019 and 20.006 mc) and one UHF channel (143.625 mc). The latter was used for conversations with ground points at a distance of about 900 miles to 1,200 miles, while the short-wave channel operated during the entire flight. Included among the radio systems was a channel for relaying flight-control commands. Its antennas were located on the clamping bands. Adjacent to it were the components of the orbital navigational system. This equipment automatically determined the spacecraft's position in space and relayed it to earth, where the data was fed into computers. On the basis of these inputs, the computers projected the spacecraft's orbital course along its flight path.

The telemetry system transmitted physiological data on the cosmonaut's condition, parameters of the ship's cabin, and data on the functioning of the ship-borne systems. Monitoring the cosmonauts' condition by telemetry was supplemented by the television system through pick-up cameras for viewing the cabin and the cosmonaut during the ship's flight over the Soviet Union.

To a large degree, the cosmonaut was protected from radiation by the command module's structure. In addition to this, he was provided with special protective chemical agents to protect him from excessive effects of radiation in case of an unexpected solar flare and high-radiation conditions. For measuring radiation levels in outer space, the ship was fitted with a dosimetric instrument, such as a Geiger counter, whose readings were automatically transmitted to earth.

During the flight, the cosmonaut wore a special suit and sat in the pilot's seat, from which he could freely make observations and pilot the ship. The pilot's couch and seat assembly were mounted on a series of gimbal rings and bearings that gave the seat 2° movement. This made it possible for the cosmonaut to maintain a horizontal attitude with respect to the earth's surface and turn 360° to position himself with respect to the sun's glare or maintain an observational position for the maximum period of time during any given orbital pass. To facilitate this freedom of movement, the equipment inside the cabin was arranged so as to give the cosmonaut the greatest freedom during the whole time of the flight either when he was secured to his seat or when he freely floated in the cabin in a state of weightlessness.

The altitude of the orbit chosen for the Vostok flights was such that it could reenter within ten days through the effects of orbital drag alone. This precaution was taken to circumvent the possibility of deorbit rocket failure. As an added precaution all vital life-support systems and food supplies were based on this ten-day contingency.

The reentry phase of the spacecraft from orbit is conducted at a predetermined time. Upon receiving a command for landing, the cosmonaut has the option to land manually or automatically. If he elects to land automatically, the ship's orientation system is switched on, which aligns the ship in space by means of a solar sensor. Then a retroengine is fired, which slows the speed of the spacecraft sufficiently (about 300 ft per sec), putting it on a reentry course. When the deorbit rockets are expended, the command module is separated from the instrument module. The thermal shield of the capsule ensures its safe reentry flight through the dense layers of the atmosphere, while the instrument module burns up there. As the command module descends, the temperature of the air boundary layer attains 10,000°F, and the aerodynamic drag on the spacecraft reaches 20 tons.

The cosmonaut can either land with the spacecraft or eject from the vehicle and land separately by means of a parachute. If the cosmonaut ejects and lands by parachute, the parachute deploys when the spacecraft reaches subsonic speeds during entry. In the first six Vostok flights, the cosmonauts ejected and landed separately from the capsule. The ejection landing procedure was as follows: At an altitude of about 23,-000 ft, the cover of the entrance hatch was automatically opened (if the cosmonaut elected to land with the spacecraft, the door of the exit hatch was not opened during the flight; the cosmonaut opened it himself after landing). Two seconds later the seat with the cosmonaut was ejected from the spacecraft by means of small rockets. The pilot's parachute system deployed immediately. At an altitude of 13,000 ft, the seat was separated from the cosmonaut, and it fell freely to earth. The cosmonaut in his pressurized space suit continued his descent by parachute. The vertical landing speed of the cosmonaut's descent is about 19 ft per sec. The same parachute that supported the cosmonaut also carried the emergency food and water rations, radio equipment, and a dinghy that would be automatically inflated in case of a landing on water.

At the same time that the cosmonaut descended by parachute, the unmanned command module returned to earth by a parachute system of its own. At an altitude of 13,000 ft, the cover of the command module's parachute container opened and a braking chute was deployed. At an altitude of about 8,000 ft, the drogue chute was jettisoned at the same time the main parachute was deployed.

The Flights

Gagarin's flight on Vostok 1 led to the conclusion that manned space flight was feasible. It showed that a person can normally endure the launching, the conditions of space flight, and return to the surface of earth. It demonstrated that the capacity of working was fully retained in conditions of weightlessness. His success also triggered the preparations for the next manned space flight.

The second manned flight was made by Titov on August 6, 1961, in Vostok 2. The orbital parameters were 152-mile apogee, 114-mile perigee, a period of 88.46 min, and an orbital inclination of $64° 56$ min. Titov's flight lasted for seventeen orbits around earth, as compared to Gagarin's single orbit. The design and equipment of Vostok 2 were similar to that of Vostok 1, but since Titov's flight was a lengthy one, special measures were taken to safeguard him from radiation exposure.

The next step in the manned exploration of space was the launching on August 11, 1962, of Vostok 3 with astronaut Andrian Nikolayev (perigee, 112 miles; apogee, 145 miles; 64 orbits) and on August 12, 1962, of Vostok 4 with astronaut Pavel Popovich (perigee, 110 miles; apogee, 146 miles; 40 orbits). The distinguishing feature of this flight was that Vostok 4 was put practically in the same orbit as Vostok 3 and in close proximity to it. The distance between the spacecrafts was only 4 miles. The flight of Vostok 3 and Vostok 4 demonstrated the solution of two key problems: the putting of two spaceships into close orbits with precision, and the establishment of two-way communication in space. The flight also showed that an astronaut retained his ability to work during a long period in a state of weightlessness, that it was possible to make a comparative analysis of the reactions of two astronauts during such a flight, and to achieve a near-simultaneous landing of two spacecraft into a preset area.

The year 1963 saw one more dual manned flight by Valery Bykovsky in Vostok 5 (perigee, 109 miles; apogee, 138 miles; 81 orbits) and by Valentina Tereshkova in Vostok 6 (perigee,

112 miles; apogee, 144 miles; 48 orbits). The mission of this dual flight was to continue the study of the effect of space flight on the human body, while the first flight by a woman provided material for ascertaining a woman's reaction.

As a result of this experiment, the conclusion was drawn that space flight was not physically detrimental to a trained man or woman. It was found that on the fourth or fifth day of weightlessness, there was a balancing of bodily reactions and adaptation to the new conditions. There remained the danger of pathological phenomena during longer flights, however, as a consequence of a possible gradual accumulation of changes in various systems and organs. This problem still called for thorough study.

The basic design of the Vostok spacecraft equipment changed from flight to flight because of unique mission requirements. Consequently, the last of the series, Vostok 6 flown by Tereshkova, was somewhat different from the first of the series, Vostok 1, flown by Gagarin three years earlier. The U.S. Gemini flights followed a comparable pattern of minor spacecraft changes with each flight.

Voskhod

The scope of the Soviet's manned space-research program increased vastly when their designers modified the single-seater Vostok into the multiseater Voskhod. This modified vehicle, the world's first three-manned spacecraft, was orbited on October 12, 1964.

The flight itself had a number of interesting features that were quite different from the earlier Vostok. For instance, the vehicle's orbit was higher than that of the Vostok ships. Its perigee was 110 miles and apogee 254 miles, as compared to about a 110-mile perigee and 150-mile apogee for the Vostoks. In this case the apogee was the critical factor. The higher the apogee, the longer it takes the spacecraft to lose speed and reenter under normal conditions. This meant that the Voskhod spacecraft could not reenter by atmosphere braking alone in case of an emergency, as was the case with the Vostok spacecraft. Therefore, to assure reentry, the Voskhod was provided with an extra deorbit system.

Moreover, in the Vostok flights, the cosmonaut had the option to either remain with the spacecraft on landing or catapult himself from the spacecraft and land separately. From a design standpoint, the former way was more practical for the multiman spacecraft design. Another Voskhod innovation was the use of solid-propellant braking rockets used during the last few feet of descent. These rockets slowed the spacecraft's vertical velocity essentially to zero as it landed. On landing the crew members could leave the spacecraft through any of three hatches.

Perhaps, the most significant design change affecting the cosmonauts was the provision for a shirtsleeve environment. Space suits were not required in the cabin under normal circumstances. Another important change was the addition of an outside television camera to observe areas that could not be seen from the porthole.

The Walk

For the four years after Yuri Gagarin flew into space, each scheduled Soviet flight solved new and complicated technical problems. One fundamental problem still remained unresolved, however, namely, man's egress from the spaceship directly into outer space. The solution of this problem was essential to future manned missions involving orbital operations such as space rescue, building of space stations and large orbital structures, as well as space repair, logistics, transfer of crews, and so forth.

To investigate the problem the Soviets modified the basic Vostok to accommodate two cosmonauts and a special air lock that gave them an entering and exiting capability when in orbit. Other Vostok support systems, equipment, and features were retained.

The air-lock chamber was attached to the outside of the spacecraft and was connected to it through a port with an airtight lid. The air-lock-chamber system was controlled by the captain from the remote control panel. The copilot could also perform the principal operations from a panel in the air-lock chamber.

The cosmonaut's egress from the air-lock chamber into space was by a porthole in the upper part of the chamber. This porthole was opened with the aid of an electrical actuator. But it could also be opened and closed by hand. The air-lock chamber had two motion picture cameras to film the cosmonauts entering and leaving the chamber, a lighting system, and the control panel. .

Outside the air-lock chamber there was a camera to record the cosmonaut's movements in outer space, air tanks for the chamber pressurization, and oxygen tanks for emergency use. The air-lock chamber was detached from the spacecraft immediately after the space-walk mission was completed.

Special space suits were developed for the space walk. These suits were a multilayer pressure suit that permitted an excess pressure to be maintained inside the suit, thereby assuring the cosmonauts' normal activities, and they were coated to shield the astronauts against excessive heat effects from solar radiation. The space suit helmets had two pressure-retained visors and it was also equipped with a light filter to protect the eyes of the cosmonauts from sunlight. Both members of the crew wore these suits, so that the captain could rescue his copilot in an emergency.

To assure the proper atmospheric environment, both aboard the spacecraft and in outer space, the space suits were ventilated, and oxygen-supply systems were provided. With both cosmonauts aboard the spacecraft, the suits were ventilated by the air inside the cabin. The moment the copilot stepped out into space and throughout his space walk, he received his oxygen supply from cylinders attached to the back of his suit.

The training of Cosmonauts P. Belayev and A. Leonov was lengthy and complicated. They first had to go through the general program as cosmonauts and then specific training for the space walk—one, as the commander of the spacecraft, the other, as copilot and the man who would actually perform the space walk experiment. Initially, the cosmonauts devoted a great deal of time studying the design of the Voskhod 2 and its systems and maintained close liaison with the designers who developed the spacecraft.

In the next phase they made extensive use of flight simulators, learning to coordinate their actions during the different phases of the flight until their responses became automatic, especially during the period of Leonov's egress into outer space and return to the spacecraft. In preparation for the walk, the cosmonauts trained aboard an airplane, in essence a flying laboratory, where they perfected all operations simulating the actual flight.

Then on March 18, 1965, the cosmonauts were launched into orbit aboard Voskhod 2 to carry out the program for which they had trained for so long. The spacecraft's orbital parameters for this flight were 108-mile perigee, 307-mile apogee, orbital inclination of 65° and period of 90.1 min.

Actually, the program for Leonov's exit began as soon as they had achieved orbit. Leonov conducted all the preparatory operations, and Belyayev ascertained that all of Leonov's life-support systems were functioning normally. At the prescribed time Belayev gave him the order to step out into space.

The walk as described by Leonov was as follows:

I climbed out of the hatch, pushed myself away from it gently, moving farther and farther away from the ship. The lifeline that connected me with the ship stretched to its full length and then my movement away from the ship ceased. The slight effort I had made in detaching myself from the ship had caused it to move slightly and I saw our wonderful spacecraft turning slowly before my eyes. I expected to see sharp contrasts of light and shadow, but there was nothing of this kind. The parts of the ship in the shadow were illuminated well enough by the sun's rays reflected from the earth. I pulled the lifeline slightly and started slowly moving away again and began to move gradually away from the spaceship, turning about my transverse axis. I saw the universe in all its grandeur. The view of untwinkling stars on a velvet black background . . . was followed by views of the earth. . . . I recognized the Volga, the mountain range of the Urals . . . as though I were swimming over a vast colorful map. . . .

I knew that it was impossible to stop my rotation by any move-ments . . . so I merely waited for the rotation to slow down when the lifeline became taut by twisting. And soon the speed of my movement gradually decreased. . . .

Some time later I made a pretty strong pull at the lifeline and was forced to protect myself from the spaceship which started moving swiftly toward me. My first thought was not to strike the spaceship with the visor of my helmet. So, as it flew toward me, I softened the blow with my hands. This proved very easy to do. . . . I felt fine, was in excellent spirits, and did not want to leave free space even after I had received the order to return to the spaceship. I pushed myself away from the hatch once more to check the origin of the angular velocity in the first moment after the push. I realized that the slightest divergence of the direction of the push produced rotation in a corresponding plane. . . . As to the so-called psychological barrier which was sup-posed to seize a man about to meet the space voids face to face, I must say that I did not feel any barrier at all and even forgot that such a barrier could exist.

During the entire flight all the systems and equipment of the Voskhod 2 functioned normally. The temperature within the cabin was approximately 65°F; the humidity, 35–40 per-cent; and the pressure 14.7 lb per sq in.

According to the flight program, they were to land during the seventeenth orbit, employing the automatic mode of de-scent and using the system of solar orientation. In the process of preparing for the landing, they noted certain abnormalities in the functioning of the solar-orientation system. As a conse-quence they had to resort to the manual method of reentry, an operation which had never been done before but was suc-cessfully performed by Belayev.

The space walk ranks equally with such epics in space re-search as the launching of the first earth satellite, Sputnik 1, and the flight of Gagarin. For their successful participation in that event, Cosmonauts Belayev and Leonov were awarded the title of Hero of the Soviet Union, and their statues, along with those of Yuri Gagarin, Sergei Korolyov, Vladimir Ko-marov, and Valentina Tereshkova were put up in Moscow. In addition, they were awarded by the U.S.S.R. Academy of

Sciences the Tsiolkovsky gold medals, instituted in honor of
the father of Soviet space travel.

Soyuz

The flight of Soyuz 1, the first of a new spacecraft series,
occurred on April 24, 1967. The spacecraft was piloted by Cosmonaut Vladimir Komarov, who previously had flown in the
multiseated Voskhod 1 on October 12, 1964. Komarov was
the first Soviet cosmonaut to go into space twice.

Soyuz 1 was injected into an orbit with a perigee of 125
miles, an apogee of 134 miles, inclination of 51.7°, and a period
of 88.6 min. After seventeen orbits the spacecraft reentered
the atmosphere normally, but the parachute failed to deploy
properly, and Komarov was killed. He became the first space
pilot, American or Soviet, to be killed in a space flight accident.
It is of interest to note that the first manned flight undertaken
without Korolyov's direction ended in the tragic death of Cosmonaut Komarov.

At the time of the accident very little information was released regarding the physical characteristics of the Soyuz
spacecraft. However, within a year after the accident, various
Soviet sources provided fairly complete descriptions of the
Soyuz, its systems, and its missions. From these accounts it
was ascertained that the Soyuz consists of three modules: the
orbital module, the command module, and the service module.

The *orbital module* resembles a miniature scientific space
station. This module is used by the cosmonauts to conduct
experiments, perform physical exercises, for recreation periods, and sleep. The orbital module also houses the control
and communication equipment (including a special all-wave
radio receiver), a portable television, and photographic, tape-
recording, and scientific equipment that normally depends
on the type of mission. For maximum lighting and observational purposes, the work station is located near one of the
four portholes.

The *command module* or pilot's cabin is directly behind and
connected to the orbital module by a hermetically sealed

hatch. The hatch, which is in the roof of the command module, permits the cosmonaut to enter before launch, and in orbit it is used for gaining access to the orbital module. The crew occupy the command module during launch, during orbital maneuvers, and during the reentry phase.

The command module contains radio communication equipment, instruments for controlling the spacecraft's descent, the life-support systems, and the food and water supply. Special containers house the main and reserve parachute systems. The spacecraft's control panel is located directly in front of the cosmonaut. The panel has instruments for controlling the ship's systems and equipment, the navigational equipment, a television screen, and control switches for the ship's on-board systems. An optical orientation device is mounted on a special porthole next to the control panel.

There are two levers for controlling the spaceship; one is mounted on either side of the chair. The right-hand lever controls the orientation in pitch, yaw, and roll; the left-hand lever changes the speed of the spacecraft while maneuvering. There are portholes on the starboard and port sides and attached to the outer shell of the spacecraft are the deorbit and the braking rockets for soft landing the spacecraft on return to earth.

The Soyuz command module is pressurized and covered by a special heat shield to protect it from the aerodynamic heating effects during reentry. Because of the outer heat shield and the inner heat-insulating material, which also serves as soundproofing, the temperature inside the cabin does not exceed 85°F.

The shape of the command module is such that it provides a limited amount of lift during its flight through the atmosphere. As a result, the reentry trajectory of the Soyuz with lift makes it possible to lower the reentry accelerations to 3–4 g, as compared to the 8–10 g experienced in the Vostok and Voskhod spacecraft. During the reentry phase, after the spacecraft is decelerated by aerodynamic drag, the braking parachute is deployed at an altitude of 5.5 miles. This is followed by deployment of the main parachute. When the spacecraft is a little more than 3 ft from the ground, the solid propellant braking motors are ignited, reducing the spacecraft's touch-down velocity to about 10 ft per sec.

The third module of the Soyuz, the *service module*, contains the main on-board equipment, the propulsion systems, and so forth, in an hermetically sealed instrument container. The container also houses the thermal-control systems, power supply, long-range radio and telemetry equipment, and instruments for the orientation and control of the spacecraft by on-board computer. The nonpressurized portion of the service module contains the liquid-propellant propulsion unit used for maneuvering the capsule and for reentry. This installation has two engines, one of which is a spare. Each engine has a thrust of 880 lb, and there is sufficient propellant to make an orbital altitude change up to 800 miles. Completing the list of equipment are the low-thrust engines for orientation and stabilization during maneuvering, and 150 sq ft of solar cells mounted on dual, deployable panels somewhat like wings on an aircraft.

The orientation and control system is one of the most important on board the spacecraft. This system orients the ship in space, stabilizes it when the engines are burning, and controls the craft when the orbit is being changed during rendezvous with another vehicle or when maneuvering near it. The system operates both automatically and manually.

Electric power is supplied to the equipment on board by a centralized electric-power system supplied by solar batteries. To ensure that the solar battery panels are constantly pointed toward the sun's rays, they are oriented toward the sun by rotating the vehicle about the axis directed toward the sun with a velocity of several degrees per sec.

Soyuz carries four television cameras; two are inside the ship's modules and two outside. The system for regeneration of the spacecraft's atmosphere is operated automatically and maintains an atmosphere inside the capsule similar to that on earth. The cosmonaut, however, can control temperature and humidity to suit his individual requirement.

The work–rest cycle in the Soyuz is based on a normal earth day. One third of the day is allotted for sleep. During the sleep

Table 3
Soviet Manned Space Launches—Second Decade*

	Soyuz 2 DUAL LAUNCH	Soyuz 3	Soyuz 4 DUAL LAUNCH	Soyuz 5
Date	10/25/68 to 10/28/68	10/26/68 to 10/30/68	1/14/69 to 1/17/69	1/15/69 to 1/18/69
Orbital altitude				
Perigee, mi.	115	113.6	107	124
Apogee, mi.	139	127.4	140	143
Total orbits flown	Approx. same as Soyuz 3	64	48	49
Flight time, hours	Approx. same as Soyuz 3	95	71 hrs. 14 min.	72 hrs. 46 min.
Orbital period, mins.	88.5	88.3	88.25	87.7
Cosmonauts	Unmanned	G. Beregovoy	V. Shatalov	B. Volynov A. Yeliseyev Y. Khrunov

	Soyuz 6	Soyuz 7 TROIKA LAUNCH	Soyuz 8
Date	10/11/69 to 10/16/69	10/12/69 to 10/17/69	10/13/69 to 10/18/69
Orbital altitude			
Perigee, mi.	115.5	128.5	130
Apogee, mi.	138.5	140.5	140
Total orbits flown	≈80	≈80	≈80
Flight time, hours	<120	<120	<120
Orbital period, mins.	88.6	88.6	88.6
Cosmonauts	G. Shonin V. Kubasov	A. Filipchencko V. Volkov V. Gorbatko	V. Shatalov A. Yeliseyev

* 1 Soyuz 1 flew on 4/23/67 in the first decade.
  2 Complete as of 12/69.

189

period, the cosmonaut rests strapped on a sofa, so as not to "float." Upon awakening, he engages in 25 min of physical exercises. Food is taken as needed, normally three to four times a day. The other periods are divided equally between work and recreation.

The Soviet's manned space program is perhaps best characterized by a series of successive phases, resolving new problems with each flight. Soviet scientists adopted this approach as the one they felt would yield the largest number of solutions in the shortest period of time. Manned Soyuz spacecraft launched in the second decade of the space age are summarized in Table 3.

13 Baikonur

> To set foot on the soil of the aster-
> oids, pick up a moonstone with
> one's hand, build stations moving
> through the ether, form living rings
> around the earth, moon, and sun,
> look upon Mars from a distance of
> a few dozen versats, land upon its
> satellites or even on the surface of
> the planet itself—all this may well
> seem more extravagant. The first
> use of jet craft, however, will open
> a new era in astronomy—an era of
> more intent study of the heavens.
>
> *K. E. Tsiolkovsky, 1911*

FOR MOST of the first ten years of space exploration, in its
public pronouncements the Soviet Union had been very care-
ful to minimize its military value. While acknowledging that
most of the astronauts were military pilots, it had not credited
other military personnel for its space achievements. Instead,
officials always asserted that space feats proved the superiority of socialism over capitalism, praising a team of "scientists,
designers, engineers, technicians and workers."

In the second decade of space exploration, however, Gen.
Vladimir F. Tolubko, First Deputy Commander-in-Chief of
the Strategic Rocket Force, said that his "soldier-rocket men"
had launched almost everything sent into space. He specifically
gave them credit for the launchings of the first artificial satellite of earth, all Soviet manned flights, the lunar series, the four
Venus probes, seven Molniya civilian-military communication
and navigation satellites, and further implied that his men had
conducted even more launches.

In order to understand some of the elements in the Soviet civilian-military aerospace complex, the organizational charts following these pages are vastly simplified so that comparisons can be made with the U.S. space organizational structure. By comparison, 40 to 50 percent of the Soviet organizational structure is similar in function to the U.S. organization.

The Soviet Communist Party Central Committee and the U.S.S.R. Council of Ministers jointly occupy the highest offices in the Soviet Union. The State Commission for Space Exploration (S.C.S.E.) administers all elements of the space program, being responsible to the Party and the Ministers. All these top positions in the Soviet space program are either held by or shared with the military. The five major, functional divisions under the Soviet space program are the Academy of Sciences, the Technical Training Organizations, the Prelaunch and Launch Organizations, and the Flight Control Organizations.

The Academy of Sciences Organizations include the Presidium, which stands at the apex of pure research activity. It is a self-perpetuating body of learned men who are responsible for the progress of scientific and scholarly work throughout the Soviet Union. Another group in the Academy is the Orbit Calculation Institute, which probably has the responsibility for running all performance and navigational type trajectories. In this regard, the Institute must establish velocity requirements for the missions, as well as determining the best flight date, transit time, and time to target. By the very nature of their duties, they must work with the Design and Flight Organizations. A third group under the Academy is the Commission for Exploration and Utilization of Space (CEUS). This organization is an outgrowth of the Permanent Interdepartmental Commission on Interplanetary Communications (ICIC) formed in 1954. Its charter is to "coordinate and direct all work concerned with solving the problems of mastering cosmic space." The present chairman is Lt. Gen. A. A. Blagonravov.

The Technical and Training Organizations of the Soviet space program are responsible for carrier-rocket and spacecraft design, mating and assembly plants; rocket-engine design and development; the spaceflight equipment plant and the cosmonaut training center.

The Prelaunch and launch-phase Control Organizations (Baikonur Cosmodrome) are the main headquarters of the State Commission; the carrier-rocket and spacecraft assembly shop; the Flight Command Center, the Operations Group of the State Commission, and the Launch Crew Unit.

The Flight Control Organizations include Flight Control Center and Guidance Stations, the Recovery Control Center, and the Telemetry and Tracking Stations.

In comparison to the Soviet space-administrative structure, the top position in the U.S. space organization is occupied by the President of the United States, and reporting to and advising the President in scientific and space matters are the Presidential Scientific Advisory Board and the National Space Council. Carrying out the U.S. space program is the National Aeronautics and Space Administration. The NASA Administrator reports directly to the President; supporting the NASA administrator are four divisions: Advanced Research and Technology, Manned Space Flights, Space Science and Applications, and Tracking Data and Acquisition.

Use of military knowledge in the Soviet space program is not limited to joint administration of the programs, the training of cosmonauts, and the use of their trained strategic rocket personnel for launch operations; it also includes the use of military facilities, specifically the specially combined ballistic missile-satellite launching facilities at Tyuratam, Kapustin Yar, and Plesetsk.

Tyuratam (*Baikonur*)

Tyuratam is the Soviet equivalent of Cape Kennedy. It is located in the heart of the empire formerly dominated by Genghis Khan, just east of the Aral Sea, near the northern part of the Syr Darya River.

The terrain surrounding the launch facility is a semiarid, flat desert valley similar to White Sands Proving Grounds

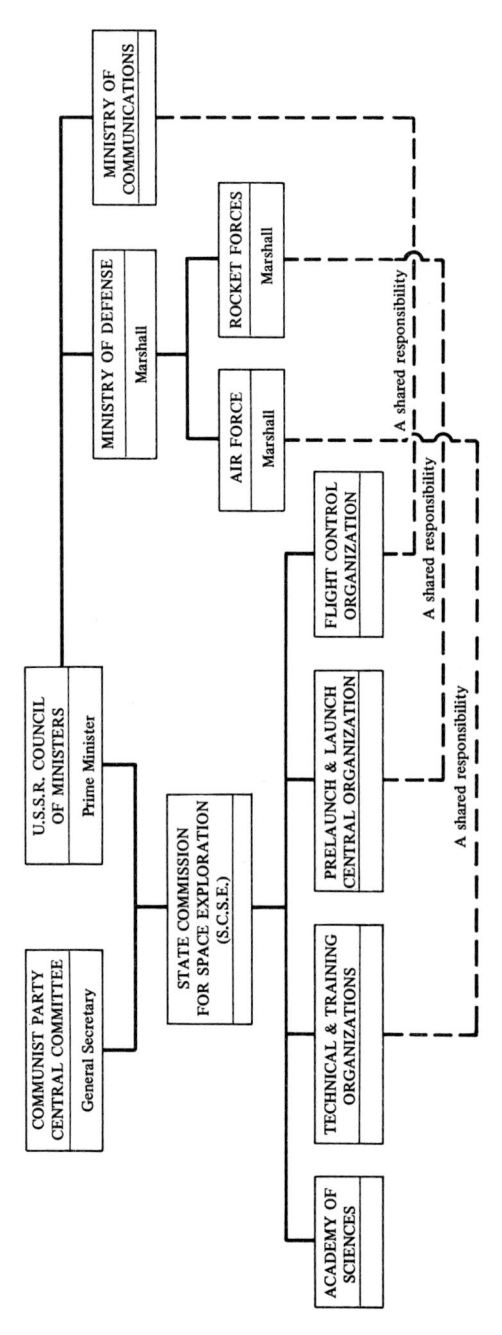

CHART VII. Soviet Space Organization, 1968

COMMUNIST PARTY CENTRAL COMMITTEE
General Secretary

U.S.S.R. COUNCIL OF MINISTERS
Prime Minister

MINISTRY OF COMMUNICATIONS

MINISTRY OF DEFENSE
Marshall

AIR FORCE
Marshall

ROCKET FORCES
Marshall

STATE COMMISSION FOR SPACE EXPLORATION (S.C.S.E.)

ACADEMY OF SCIENCES

TECHNICAL & TRAINING ORGANIZATIONS

PRELAUNCH & LAUNCH CENTRAL ORGANIZATION

FLIGHT CONTROL ORGANIZATION

A shared responsibility

A shared responsibility

A shared responsibility

A shared responsibility

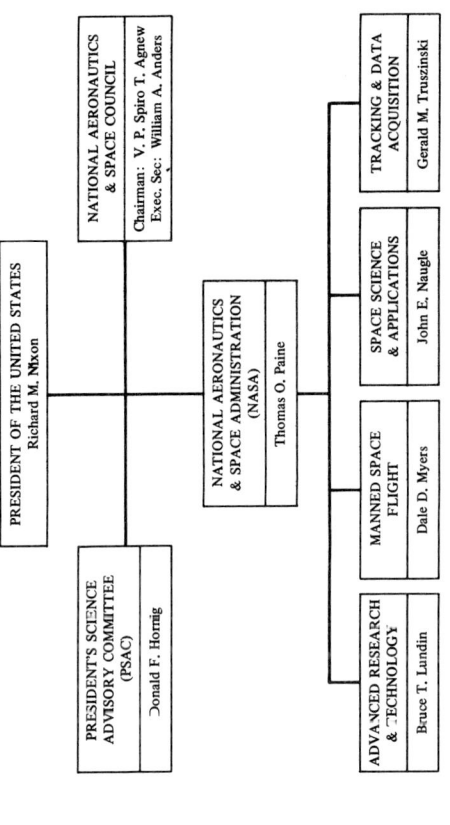

PRESIDENT OF THE UNITED STATES
Richard M. Nixon

NATIONAL AERONAUTICS & SPACE COUNCIL
Chairman: V. P. Spiro T. Agnew
Exec. Sec: William A. Anders

PRESIDENT'S SCIENCE ADVISORY COMMITTEE (PSAC)
Donald F. Hornig

NATIONAL AERONAUTICS & SPACE ADMINISTRATION (NASA)
Thomas O. Paine

ADVANCED RESEARCH & TECHNOLOGY
Bruce T. Lundin

MANNED SPACE FLIGHT
Dale D. Myers

SPACE SCIENCE & APPLICATIONS
John E. Naugle

TRACKING & DATA ACQUISITION
Gerald M. Truszinski

CHART VIII.   U.S. Space Organization

(W.S.P.G.). Like the W.S.P.G., one of the major advantages of Baikonur is that the range passes over vast uninhabited areas, a safeguard in the event of an abortive flight.

The isolation of the Baikonur facility contributes to maintaining the secrecy that surrounds the launches of Soviet carrier rockets and spacecraft. Baikonur is also used as a research and development and operational flight center, with emphasis on sensitive military launches. From 1957 to 1967, 73 percent of all the Soviet space launches originated from Baikonur: all the early Sputnik launches; the first Mars, Venus, and lunar probes; the first manned prototypes; and the later Vostok, Voskhod, and Soyuz manned flights. Baikonur also handles all the heavy Cosmos- and Elektron-type scientific satellites, the communication and meteorological satellites, and such heavy (over 10,000 lb) military satellites as the Vostok-type reconnaissance and fractional-orbital-bomb (FOB) weapon system.

For its contribution to the early development of unmanned and manned space flights, the Soviets have honored Baikonur with a marker as an historical site.

Kapustin Yar (K.Y.)

With the growth in Soviet space activities, a simpler and less busy facility was needed where scientists could be closer to the launch site. Kapustin Yar, the oldest and perhaps the largest ICBM installation in Russia, was chosen and made into a combined ICBM/satellite launching site. It is also one of the sites where some German scientists from Peenemunde were put to work after World War II.

K.Y. is situated in the low-lying Volga River Basin, sixty miles east-southeast of Stalingrad (now Volgograd). From this site, payloads up to 2,000 lb are launched eastward over the Caspian Sea. K.Y. is believed to have carried the main load of the early Soviet atmospheric-research rocket launchings. More recently 16.6 percent of all the Soviet orbital launches originated from this site.

When the first Cosmos was launched, the Soviets announced that several cosmodromes would be used. Since then, essen-

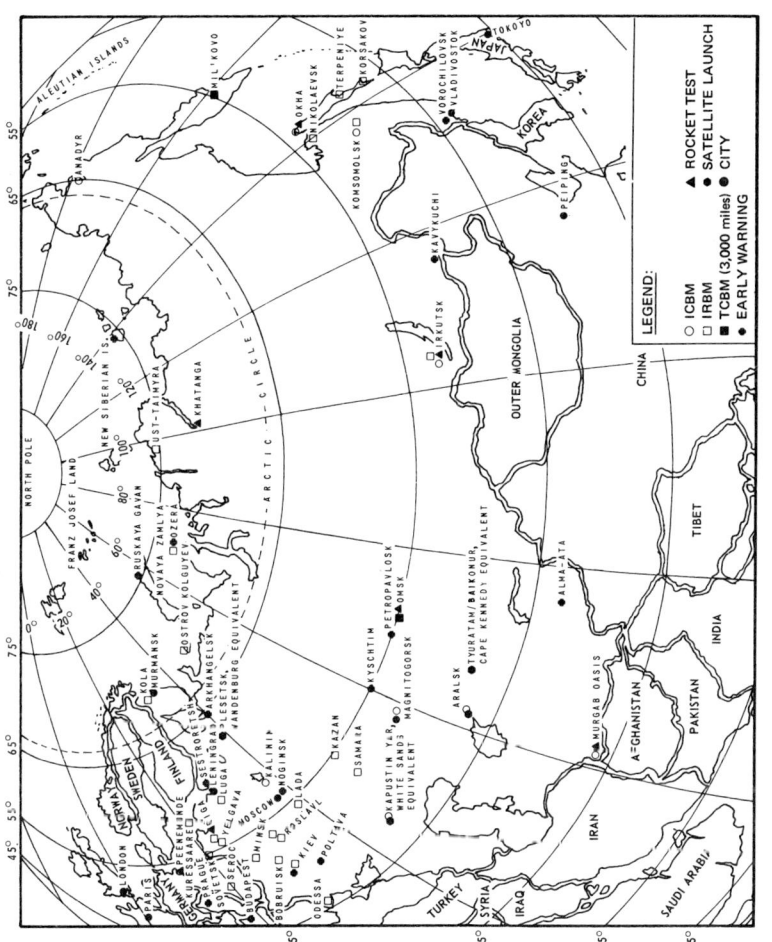

CHART IX.   Soviet Launch Bases

tially all the launches from K.Y. were Cosmos types. This site is compared to Cape Kennedy in that rockets launched from K.Y. can be seen by passengers aboard ships traveling on the Caspian Sea. Unlike Cape Kennedy, however, the K.Y. launches pass over several populated areas.

Plesetsk

With the enormous development of spacecraft technology, a site was needed for the more specialized satellite. Some of these satellites required daily worldwide coverage that could only be obtained by polar or high-inclination orbits.

Table 4
Satellite Launches by Launch Site*

Year	Tyuratam	Kapustin Yar	Plesetsk	Launches per year
1957	2	0	0	2
1958	1	0	0	1
1959	3	0	0	3
1960	5	0	0	5
1961	6	0	0	6
1962	13	7	0	20
1963	13	4	0	17
1964	23	7	0	30
1965	41	7	0	48
1966	31	7	6	44
1967	33	8	25	66
1968	35	9	30	74
1969	30	5	35	70
Totals	**236**	**54**	**96**	**386**
Percent of total	61.1	14.0	24.9	

* Compiled by U.S. sources. Does not include multiple payloads

Table 5
Major Nuclear Missile Delivery Systems

Name	Propellant	Range (statute miles)	In service	Estimated warhead
U.S.				
LGM-25C Titan II	SL	12,000	1963	5+ megaton
LGM-30A Minuteman 1	S	6,300	1962	1+ meg.
LGM-30F Minuteman 2	S	9,200	1966	2 meg.
UGM-27A Polaris A1	S	1,380	1960	0.7 meg.
UGM-27B Polaris A2	S	1,700	1963	0.7 meg.
UGM-27C Polaris A3	S	2,850	1964	0.7 meg.
MGM-13B Mace	T	1,380	1963	kiloton range
MGM-31A Pershing	S	400	1964	kiloton range
MGM-29A Sergeant	S	75	1962	kiloton range
U.S.S.R.				
ICBM Scrag	L	Orbital	†	?30 megaton
ICBM*	L	5,000+	1963	10 meg.
ICBM*	SL	5,000	1963	10 meg.
ICBM Sasin	SL	5,000+	1963	10 meg.
ICBM*	SL	5,000+	1965	10 meg.
MRBM Shyster	L	750	1956	kiloton range
IRBM Skean	L	2,100	1961	1–2 meg.
MRBM Sandal	L	1,100	1959	1 meg.
SSLM Sark	SL	400	1959	1 meg.
SSLM Serb	SL	650	1961	1 meg.
Cruise SLM*	SL	400	1961	kiloton range
SRM Scud	L	70	1957	kiloton range
CRM Shaddock	T	250	1961	kiloton range

Legend

L	Liquid fuel	SLM	Ship-launched missile
S	Solid fuel	SSLM	Submarine- and ship-launched missile
SL	Storable liquid fuel		
T	Turbojet	SRM	Short-range missile
†	Believed not operational	CRM	Cruise missile
		*	No name assigned

Note Rockets claimed to be a solid-fueled ICBM (Savage) and a mobile MRBM/ICBM (Scamp) have been shown in Moscow parades, but are not believed to be in service.

For this purpose, the Soviets converted the Archangel missile site, near the town of Plesetsk, into the third combined ICBM/satellite site. Since 1966, this site has been used for space launches. Since then many weather, navigation, high-inclination scientific, and military characteristic satellites have been launched from the base.

The Plesetsk site, still unidentified by the Soviets, is comparable to the U.S. West Coast Vandenberg Air Force Base and Point Arguello facilities in application and function, namely, for military polar launches.

Other Facilities

Besides the three major space-launching facilities, the Soviets have extensively used Heiss Island in the Franz Josef archipelago in the Arctic Ocean and Mirny in the Antarctic for upper atmospheric research launch sites. These sites correspond to NASA's Wallops Island. There were fifty-eight launchings during the I.G.Y. and I.G.C. from these two bases.

Military Systems

No historical reference of the Soviets' space capability can be complete without at least a brief mention of their military arsenal. Some of the military ballistic systems that have come to light in recent years that comprise in part General Tolubko's arsenal are shown tabulated on the preceding page, with a comparison of U.S. system.

14 A Decade of Accomplishments

> Everything of which I speak is
> merely a feeble attempt to foresee
> the future of aviation, aeronautics,
> and rocketry. In one thing, I firmly
> believe that the Soviet Union will
> be first.
>
> *K. E. Tsiolkovsky, 1935*

IN SPITE OF Russia's early space superiority, the United States now has a commanding lead in many categories of space flight, as witness the moon walks. The U.S. has had overwhelming success with large boosters, manned space flight, satellite technology, application satellites, and have contributed to the fields of space science and basic research.

By the end of 1967, the United States' Saturn 5 space booster had more than twice the lift-off thrust and ten times more payload carrying ability than the Soviet Union's Proton. All told, U.S. space boosters successfully launched 388 payloads into orbit compared to 239 for the U.S.S.R.'s space boosters. The success ratio for the ten years of cumulative activity is 91 percent for the United States and 87 percent for the Soviet. These launches resulted in twenty-five firsts for the United States as compared to twelve for the Soviet Union.

In manned flight the United States has had sixteen successes, launching two men into a suborbital trajectory and twenty-four men (many of them twice) into orbital flights. These flights have credited the United States with the highest altitude attained by man (851 miles), the longest time in orbit (330.35 hr), the longest single and total extravehicular activity

(EVA) records, 5 hr 58 min and 12 hr 39 min. While accomplishing these feats, American astronauts have logged 1,993.37 manhours in orbit compared to 532.53 manhours for the Soviet cosmonauts and established eleven significant world records compared to three for the Soviet Union. The astronauts conducted ten rendezvous and nine physical docking flights; while docked, they used for the first time both the docking vehicles' propulsion and propellants to change orbital attitude and inclinations.

Thus, it was during the Gemini program that the United States really overtook and surpassed the Russians in the manned space effort and all of the flight requirements for Apollo were proved—rendezvous, docking, EVA, and physical endurance in zero gravity. Gemini was the program in which all of the key Apollo commanders gained their initial spaceflight experience.

Subsequently, during the flights of Apollo 8 and Apollo 10 six of our astronauts orbited the moon, paving the way for Apollo 11 and man's first landing on the moon.

At 10:56 P.M. on Saturday, July 20, 1969, Astronaut Neil Armstrong placed his foot firmly on the fine-grained lunar surface. Pausing briefly, he spoke the now famous words, "That's one small step for a man, one giant leap for mankind." These words brought to a close a chapter in American history first set by President John F. Kennedy in May, 1961, when he boldly committed his country to land men on the moon before 1970.

International Achievements

Space, which in the 1960s was predominately the concern of only two nations, will in the latter part of the 1970 decade manifest itself on an international scale. Before multinational space ventures become a reality, however, the participating nations will have to pace their progress with the development of international space law.

Many precedents, such as the I.G.Y. and I.C.Y., for cooperative space efforts exist. There are also many international groups in which future cooperative efforts might develop: the charter of the United Nations Committee on the Scientific Uses of Outer Space, the European Launched Development Organization (ELDO), the European Space Research Organization (ESRO), and the Western Meteorological Organization (WMO).

The United Nations has produced two significant agreements: U.N. Resolution 2222 (XXI), entitled *Treaty on Principles Governing the Activities of States in the Exploration and Use of Outer Space, Including the Moon and Other Celestial Bodies,* and Article 2, from U.N. Resolution 2345 (XXII), *Agreement on the Rescue of Astronauts, the Return of Astronauts, and the Return of Objects Launched into Outer Space.* The United States in turn has space agreements in force between itself and sixty-five nations. Futhermore, NASA's Office of International Programs is well established and supported by the State Department and the National Academy of Sciences. A total of ninety-nine nations are participating in NASA's International Programs.

The Soviet activities in international space cooperation have been comparable in scope to the United States'. Soviet space scientists have worked with scientists of Bulgaria, Hungary, the German Democratic Republic, Cuba, Mongolia, Poland, Rumania, Czechoslovakia, and other socialist countries. The area of cooperation with most of these countries is the visual and photographic observations of artificial earth satellites, which began as far back as 1957 when Sputnik 1 was launched.

The experience accumulated made it possible to change over, beginning with 1962, to multilateral cooperation in observing satellites, thus making it possible to start more complicated research programs with the help of observers in many countries. Besides scientists of the socialist countries, those of the Netherlands, Greece, Italy, Finland, Sweden, and other nations are participating in these programs. Through the joint efforts of a number of countries, synchronic photographic observations have been made for geodetic purposes, as well as tracking of low flying satellites for the study of transitory changes in the density of the atmosphere. About twenty countries are regularly sending the Astronomical Council of the U.S.S.R. of Science the results of observations of man-made space objects.

The international bulletin *Observations of Artificial Earth Satellites* has been published since 1963. It carries the reports of the cooperating countries and articles on the most interesting research work. Researchers taking part in the effort gather annually at scientific conferences held in different countries. Sessions of the specially organized Commission for Research by Means of Observations of Artificial Earth Satellites are also held regularly. A school for young observers of artificial earth satellites was organized in Tashkent in 1965. Forty students from eight socialist countries are being trained there under the guidance of scientists.

The Soviet Union's cooperation with a number of nonsocialist countries in space exploration is also progressing. The first practical results of this cooperation are the Soviet-French television experiments. For several years Soviet and French physicists have been engaged in a joint study of the complex of electromagnetic phenomena in magnetically conjugate points of the earth. Soviet scientists have worked on the Kerguelen Islands in the Indian Ocean, and their French colleagues in the village of Sogra, Archangelsk region.

As a result of the signing of the Soviet-French Agreement on Cooperation in the Study and Exploration of Outer Space for Peaceful Purposes, more cooperation between Soviet and French scientists is expected. Under this agreement, Soviet-French cooperation in space exploration includes the launching of a French satellite by the Soviet Union. Also, included in the program is space meteorology research and the study of space communications via artificial earth satellites.

For several years now Soviet scientists have cooperated with the staff of Jodrell Bank, the well-known British radio astronomical observatory. The Soviet Union, along with some other countries, is assisting India to build a rocket range for launching geophysical rockets in the area of the geomagnetic equator, which is at an equal distance from the earth's magnetic poles. For this purpose the Soviet Union has presented India with a helicopter, an electronic computer, and some other research equipment.

Some Afro-Asian countries have either built or are building, with Soviet aid, stations for the observation of artificial earth satellites and other space objects. The aid given for the organization of such stations and the training of local personnel enable an increasing number of countries, which so far cannot use rockets and satellites for their own research, to contribute to the joint effort of studying and exploring outer space.

There is also bilateral space-research cooperation between the Soviet Union and the United States, under a special agreement concluded on June 8, 1962. The agreement stipulates that cooperation is to be, primarily, in three fields: the employment of artificial earth satellites for meteorological purposes; for the organization of space communications; and for compiling a magnetic map of the earth and developing the science of terrestrial magnetism.

As regards meteorology, it has been decided to have an extensive exchange of meteorological information, both conventional and that received from satellites. It is contemplated that both sides will improve their experimental meteorological satellites so that in the future they will be able to go over to coordinated launchings of operational meteorological satellites with a rapid dissemination of the data obtained.

World meteorological centers have been set up in Moscow and Washington, and a direct communication line has been established between them. The line, which became operational in 1964, is used for round-the-clock transmission of the most important information on the state of the atmosphere. The Soviet Union and the United States have come to an understanding that in the future all other interested states will be able to use the line for their needs.

The first American-Soviet experiment in the field of communication satellites was the establishment of communications, via space, by means of the American satellite Echo 2. In the spring of 1964, this satellite was used for thirty-four communication sessions between Jodrell Bank and the Gorky University

Observatory in Zimenki. Besides these three lines in Soviet-American cooperation, an understanding was reached on the joint preparation and publication of a work summarizing the main achievements in space biology and medicine during the past decade and showing the prospects of their development.

Humanitarian Achievement

The technological achievement of landing men on the moon in this decade is perhaps comparable to the giant leap forward for mankind exemplified by the actions of the Apollo 11 astronauts when they left behind decorations awarded by the Soviet government to two of their cosmonauts who have died. Yuri Gagarin, the first man to orbit earth, later died in an airplane crash and Vladimir Komarov was killed when his Soyuz spacecraft plunged to earth. Their widows gave the medals to Frank Borman, Commander of Apollo 8 (which was the first manned lunar-orbiting flight) while he was visiting the Soviet Union in July, 1969. Borman, in turn, passed the Soviet medals on to the Apollo 11 crew, who left them on the moon.

15 A Projection of Soviet Space Activities

> It may be that hundreds of years
> will pass before my ideas are actu-
> ally realized and mankind makes
> use of them to extend its control
> beyond the face of the earth to
> include the entire universe.
>
> *K. E. Tsiolkovsky, 1911*

THE ADVANCE of civilization is paced by man's quest for new frontiers to conquer. Today, man's frontier is space. Space, however, unlike earlier explorations, is a unique undertaking: it is imperialism without geographical reward, colonization without natives, military without massive confrontation, and technology with overwhelming political overtones. The world watches as the United States and the Soviet Union commit vast portions of their financial, technological, and industrial resources to send carefully selected and trained men into this new and hazardous environment.

It is quite reasonable to expect that the momentum generated by the Soviet Union in the first decade of space will carry it into the next decade with determination. The question, therefore, is not will the Soviets continue, but in what direction and with how much emphasis. There are significant statements upon which to base an estimate of future Soviet activity. The reliability of these statements from key individuals within the Soviet Union is best judged in light of remarks made concerning Sputnik before its first flight. Since then, comparable statements on the future of Soviet space activities have appeared in various Soviet publications.

Predictions About Soviet Space Science

The directives of the Twenty-Third Congress of the Communist Party, U.S.S.R., on the Five-Year Plan for Development of the National Economy for 1966–1970 provide for the further study of space and the use of the results for improvement of radio communication, radio navigation and television, meteorology, and so forth. The Soviet communications satellite, Molniya, has laid the beginning of a system of distant communications and other important applications of artificial earth satellites for solution of many terrestrial problems. Equipping satellites with superior electronic apparatus will lead to improved sea and air navigation. A vessel at any point in the ocean will be able to interrogate a satellite and obtain precise data on the position and velocity of motion of the satellite at this particular time. Using data on satellite motion and its orbital parameters, the navigator will be able to compute precisely the positions of the vessel in the ocean. Satellites will greatly improve the field of meteorology. In the near future, such satellites will create a global observation system. Study of processes occurring on the sun will result in improved forecasts of the propagation of radio waves in earth's atmosphere, since the state of the latter is essentially dependent on solar activity. For improvement of radio communication via artificial satellites, in addition to active systems such as Molniya 1, there may be systems of passive and semipassive types. In these systems, it is effective to use antennas which have the capability of reflecting radio waves in the direction from which they are received. Also of great promise are the methods of quantum electronics, especially lasers and semiconductor instruments.

Predictions About Television

An item on a page titled "News and Briefs" in an East German radio magazine says that Professor Mstislav V. Keldysh,

President of the Academy of Sciences, U.S.S.R., told the general assembly of the Academy that within five years the Soviet Union will have solved the technical problems involved in receiving satellite relayed television broadcasts with ordinary commercial sets without the use of intermediate transmitters.

Manned Space Stations and the Future

I. N. Bubnov and L. N. Kamanin state in *Manned Space Stations:*

> What are the main stages in the difficult conquest of space? Very different forecasts on this matter are often encountered in the foreign press. Some of them are unjustifiably optimistic, while others are too pessimistic. Nevertheless, it is possible to give an approximate succession of further events in space conquest:
> 1. A soft landing of an automatic station in 1964–1965;
> 2. Landing of a man on the moon in 1968–1970;
> 3. Building of a larger manned space station with a crew of thirty to fifty men in 1972–1975;
> 4. Flights to Mars and Venus of a manned spaceship with return to earth in 1975–1980;
> 5. Landing of men on Mars in 1980–1990.

Predictions About the Moon

N. N. Semenov, Vice President of the U.S.S.R. Academy of Sciences and Nobel Prize winner, considers the soft landing achieved by the Soviet automatic lunar station the outstanding scientific event of recent years. He is firmly convinced that new discoveries very soon will lead to the exploration and exploitation of the moon by man. The question naturally arises as to what attraction there can be in such a wasteland, devoid of water and air. Semenov answers by saying that the moon, because of these very properties, is ideally suited to the building of powerful electric stations tranforming the sun's rays into electricity and other forms of energy. Going further he envisages the moon as a grandiose power station for earth. The problem will be the transfer of this energy to earth. For this,

Semenov suggests the possibility that modern masers, giving
an exceptionally narrow and highly concentrated beam of
radio waves, may with time be so improved that they can
efficiently transmit energy from the moon to earth.

Future Space Exploration

In commenting on the launching of the Voskhod, Leonid
Byshko makes the following comments on the future prospects
of space exploration.

> After conquest of the moon . . . there will be flights of
> manned ships to the planets of the solar system. This will re-
> quire new carrier rockets based on the use of nuclear energy.
> . . . The only requirement is time to design, build, and perfect
> them. The further penetration of man into space will be accom-
> panied by improvement of spacecraft for flights on near-earth
> trajectories. Such craft can be used as rapid transportation. For
> example, flight time from Moscow to New York on a passenger
> spacecraft will not exceed one-half hour, and their maneuvera-
> bility will make it possible for them to land in any specified place,
> including in densely populated places.

In *Ten Years of Soviet Space Research*, M. K. Tikhonravov
and his associates predict, "The time has come to carry out
projects which once seemed fantastic, to build research stations away from earth, to send men to Mars, Venus, and the
other planets."

Achieving the stated goals requires an immense planning
and technological activity that could span a decade as witnessed by the U.S. Apollo program. These flights will not be
possible without maximum effort.

Accurate flight timing is essential because space flight is
still restricted by the capabilities of propulsion systems. This
affects flying into earth orbits, making earth orbital changes,
flights to the moon, and especially flights to the planets, and
limits the U.S. to the use of minimum energy trajectories. From
an energy standpoint orbital flights are the only ones that can
be flown at any hour of the day, every day of the year. More
ambitious flights are restricted by launch windows.

Probable Future Use
of Lunar and Planetary Windows

Lunar launch windows are the most frequent, occurring on a monthly basis. The only restrictions on lunar flights are generally predicated on the mission or communication requirements.

Least-energy launch opportunities for flights to the planets occur only during those time intervals when the relative positions of the earth and the target planet are such that the spacecraft reaches the target on roughly the side of the sun opposite from where it left earth. These favorable launch opportunities recur at intervals that may be approximated by the synodic period (the time interval between identical relative positions of the two planets).

Because of national commitments, budget availability, and state of technical development, it is reasonable to expect that some planetary windows will not be used for flights and that the less frequent planetary flight windows will receive the greatest consideration. It should be noted that the Soviet has used all windows except one since its planetary program started.

The Future of Launch Vehicles

The standard Vostok launch vehicle in its various staged configurations will be phased out as an active space booster. The Proton launch vehicle will come of age in 1970s. Its role in the Soviet space program and its payload capability will be better defined. The Lunar launch vehicle will have its initial flight by the end of the 1970s. The unveiling of the LLV booster is directly related to the date of the Soviet's first manned lunar landing.

A prototype nuclear stage will be developed between 1972 and 1975. The incentives to develop the nuclear stage are lunar logistics and planetary travel. Information on the Soviet version of a lunar spacecraft and its modules will begin appearing in the press sometime in mid 1970s. Soviet literature is replete

Table 6
Solar System Flight Windows

Planetary targets	1970	1971	1972	1973	1974	1975	1976	1977	1978	1979	1980
Moon	X	X	X	X	X	X	X	X	X	X	X
Mercury, Jupiter, Saturn, Uranus, Neptune, & Pluto	X	X	X	X	X	X	X	X	X	X	X
Mars		X		X		X		X		X	
Venus	X		X	X		X		X	X		X
Venus–Mercury flyby	X			X					X		
Venus–Mars flyby				X		X			X		
Jupiter–Saturn flyby								X			
Earth–Jupiter–Saturn–Uranus–Neptune Grand Tour									X		
Asteroid belt								X			
Comet Encke flyby					X						
Comet Tempel flyby			X								
Pons–Winnecke flyby	X						X				
Schwassmann–Wachmann flyby					X						

Notes on multiplanetary and comet flight windows: 1) Venus–Mercury and Venus–Mars windows occur three times in a decade; 2) Jupiter–Saturn window available in 1977 and occurs once in a decade; 3) Grand Tour—Earth, Jupiter, Saturn, Uranus and Neptune window available in 1977 and occurs once in 183 years; 4) Asteroid belt and Comet windows basically occur once in a decade. Multiplanetary, asteroid and comet flights are of extreme importance both to the scientific and space communities. To date, however, no flights to these targets have been scheduled. But because of the scarcity of available windows, it is reasonable to assume that the United States and the Soviet Union will in the near future plan flights to one or more of these planetary bodies.

Table 7
Manned Lunar Exploration

(projection)

Year	United States	U.S.S.R.
1970	Two Apollo landing flights	—
1971	Two Apollo landing flights	—
1972	—	First Soviet manned landing on moon
1973	Two Apollo landing flights	Soviet landing flights continue
1974	Last single Apollo landing flight	Soviet landing flights continue
1975	—	Soviet landing flights continue
1976	Begin limited lunar base construction	Begin limited lunar base construction
1977	Base construction continues	Base construction continues
1978	Base semi-operational 3–6 men for 2 weeks	Base semi-operational 3–6 men for 2 weeks
1979	Base semi-operational 3–6 men for 30 days	Base semi-operational 3–6 men for 30 days
1980	Base fully operational 10–15 men for 3 months	Base fully operational 10–15 men for 3 months

with information on how Soviet Russia will use the moon for exploitation and scientific purposes.

In the period between 1972 and 1975, lunar exploration events will be shared with the construction and building of space stations in earth orbit. To support Soviet logistics associated with these operations, recoverable and reusable earth-orbital systems will be introduced in the 1972–1975 period.

Manned Earth Orbital Space Stations

The early space station (1970–1973) will be quasimilitary, capable of supporting a few men in orbit for a period of about a month. Later space-station generations will be primarily military, capable of supporting up to twenty-four men in orbit for perhaps as long as three to six months. The eventual evolution of the space station (1975–1978) will be military and will support about fifty men in orbit for at least six months.

Table 8

Space Station Construction

(projection)

Year	United States	U.S.S.R.
1970	—	Earth orbital station, 3–9 men for 30 days
1971	—	—
1972	Apollo applications station 3 men, 14–30 days	—
1973	—	—
1974	—	—
1975	Semipermanent station; 12 men in orbit for 90 days; station utilizes recoverable and reusable logistic system	Semipermanent station; 24 men in orbit for a minimum of 90 days
1976	—	—
1977	—	—
1978	Very large permanent station begins—approximately 50 men in orbit for 3–6 mos.	Very large permanent station begins; approximately 50 men in orbit for 3–6 mos.; utilizes recoverable & reusable logistic system
1979	—	—
1980	—	—

Earth Orbital Applications Satellites

Extensive use of near-earth applications satellites, communications, weather, television, and earth resources, will be highly accelerated throughout the 1970s. Applications satellites with a long-term orbital capability will begin to make their appearance in the 1970s.

Most logical candidates for integration into a multipurpose satellites are the navigation, communication, meteorological, and earth-resources satellites. The ultimate evolution of applications satellites is the integration of many of their functions into the manned space station.

Unmanned Planetary

Typically, the information to be sought during the earliest windows will be of the type that accrues the greatest amount of scientific prestige and also seeks information pertinent to critical long-lead items associated with the possibility of manned flights.

For future planetary windows, one should expect to see the continuance of the multiple-launch philosophy, with each suc-

Table 9

Planetary Exploration

(projection)

Year	United States	U.S.S.R.
1970	—	—
1971	Mars flyby	Mars probe
1972	Pioneer F—Jupiter probe	—
1973	Mars probe Venus–Mercury flyby Pioneer G—Jupiter probe	Mars flyby
1974	—	—
1975	Mars orbiter/lander	Mars flyby
1976	—	—
1977	Mars orbiter/lander	Mars orbiter/lander
1978	—	—
1979	Mars rover	Mars rover
1980	—	—

ceeding launch being more sophisticated than the previous launched orbiters and landers. As such, one can expect an orbiter with television and a lander. The ultimate aim of the unmanned precursor flights is to gather sufficient information on interplanetary space and the planets so that manned flights can follow.

At the end of the 1970s, the Soviets will probably assemble and launch the first interplanetary manned expedition to Mars. In the Soviet Union, space stations and nuclear propulsion are synonymous with manned planetary travel.

Table 10
Projected Soviet Space Program Gross Cost Estimate

Projected mission	Millions of dollars
Lunar exploration	$ 1,000
Space station, 3–5 men	3,000
Space station, 35 men	15,000
Space station, 50 men	7,000
Mars lander	225
Venus lander	225
Mars orbiter/lander (2)	400
Venus orbiter/lander (2)	400
Venus orbiter/lander (2)	400
Mars manned lander	2,400
Mars-Venus flyby	325
Jupiter-Saturn flyby	325
Grand tour	300
Total	31,000
Average over 10 years	3,100
Supporting activities	1,400
Average annual expenditures	4,500

Gross National Product and Space Budgets

Aerospace, like other technological undertakings, are generally budget constrained. Therefore another check on the validity of the projected programs is to estimate their cost and to compare it against past funding rates.

The Soviet G.N.P. since 1950 has been approximately half of the U.S. G.N.P., based on equivalent U.S. currency. Yet the Soviet space budget on a fiscal basis has always been comparable to or slightly larger than the U.S. space budget. This im-

Table 11

Comparison of Gross National Product and Total Space Expenditures

| | United States | | U.S.S.R. | |
| | GNP | Space | GNP | |
Fiscal Year	(in billions of dollars)	Expenditures (in percent)	(in billions of dollars)	Space Expenditures*
1957	$450	0.150%	$210	
1958	450	0.248	240	
1959	480	0.433	252	
1960	504	0.888	262	
1961	520	1.467	278	
1962	560	2.386	288	
1963	590	4.078	294	
1964	632	5.929	317	
1965	684	6.886	335	
1966	750	7.718	358	
1967	794	7.237	375	
1968	866	6.666	396	
1969	930 est.	6.325	407 est.	

* United States sources estimate Soviet space expenditures to be comparable to U.S. expenditures, on a fiscal basis.

plies that the Soviet allocates at least twice the percentage of its G.N.P. to space that the U.S. allocates (NASA plus DOD), in order to maintain a space budget equivalent to or greater than the U.S.

From the estimate of the projected Soviet space programs, it appears that the U.S.S.R. programs will be financed at a rate that is equivalent to the NASA expenditure over the last few years. In 1967, the United States spent approximately 0.83 percent of its G.N.P. on NASA and DOD space-related activities. To finance the projected Soviet programs would require about 1.2 percent of the Russian G.N.P. From these percentages, one can conclude that the outlined U.S.S.R. space program is economically feasible.

EPILOGUE:
Perspectives

> We can confidently state that the scientific legacy of K. E. Tsiolkovsky will not be preserved in the manner in which archivists preserve ancient writings. Soviet science is multiplying the scientific achievements of the founder of rocket dynamics. The revolutionary sweep and the Bolshevist spirit of determination in the solution of new problems which the Communist party has instilled into the Soviet people are a guarantee of the successful materialization of Tsiolkovsky's most audacious dreams.
>
> *A. A. Blagonravov, Academician,*
> *1954*

WHEN one looks back at the long history of rocketry and space travel, it almost becomes axiomatic that nothing can keep man from exploring space. The trend in history is unmistakably toward interplanetary travel.

As the Soviets See It

History tells us that the first rockets were built in China more than a thousand years ago. Nevertheless, none of those rocket builders, none of the millions of people who witnessed fireworks and pyrotechnic displays, created a new science of the theory of rocket flight. Furthermore, gunpowder rockets had attracted the attention of large numbers of highly competent military specialists in the mid-nineteenth century. Despite this, prior to Tsiolkovsky, there was no theory of reaction propulsion.

Having derived the initial working formulas of rocket motion, Tsiolkovsky outlined a broad program for successive advances in rocketry. The principal aspects of his program were experiments in situ (meaning laboratories where static tests are performed); plane motion of a reaction machine (on an airfield); ascents to a low altitude and gliding descents; penetration to the more rarefied layers of the atmosphere, such as the stratosphere; flights beyond the confines of the atmosphere and gliding descents; establishment of independent space stations outside the atmosphere. Utilization of solar energy by astronauts for respiration, nutrition, and certain other physiological purposes; utilization of solar energy for industry and for travel throughout the solar system; visits to the smallest bodies of the solar system and the colonization of man throughout the solar system.

As President Richard M. Nixon Sees It

In a post-launch speech to the Apollo 12 team in the fire control room at the Kennedy Space Center, President Nixon said:

> America, the United States, is first in space. We're proud to be first in space. We don't say that in any jingoistic way. We say it because as Americans we want to give the people of this country, and particularly our young people, the feeling that here is an area on which we can concentrate for a positive goal, concentrate and be proud of being Americans. Be proud of what we have accomplished, not only for ourselves, but for future generations and for the whole world.

As the Author Sees It

Aside from politically and militarily exploiting their space successes, the Soviets have encouraged the growth of a very capable and large scientific community. The strong technical base established in the past few decades will permit them to concentrate on orbital, lunar, and planetary areas of conquest. This will be done by utilizing the most sophisticated techniques and equipment such as recoverable and nuclear stages in conjunction with spacecraft capable of operating in a spatial and atmospheric environment.

Consequently, with a good press, planned publicity, a strong technological base, national support, and adequate budget, the Soviets will continue to exploit their scientific capabilities for political prestige, for the national ego, and for the historical record.

As Vice Admiral Hyman G. Rickover Sees It

The inferences of the scientific and technological advances made by the Soviets was commented on by Vice Admiral Rickover in 1961.

If we are to take intelligent and resolute action that will ensure survival, we must have a clear conception of what the contest is about. It is difficult, however, to muster the necessary detachment and breadth of view since this is a conflict between two diametrically opposed types of society, and we are personally identified with one of them. The temptation is great to see it as nothing more complicated than evil bent on destroying good. Yet often Russian achievements are not specifically communistic or undemocratic, simply efficient. We only harm ourselves if we close our eyes and ears to this fact. It is realistic, not disloyal, to recognize that at times the Russians have devised a better method than we. Also, when we dismiss their successes by dwelling fondly on the admitted fact that life is sweet here and bitter and harsh under the totalitarian yoke, our argument is irrelevant. The United States and Russia are not engaged in a popularity contest, but in a grim technological race, with the survival of freedom in this world at stake.

APPENDIX A
Space Launches– (1957–1969)

Appendix A*   Space Launches—(1957-1969)

Satellite name	Date launched	Launch site**	Apogee (kilometers)	Perigee (kilometers)	Inclination (degrees)	Period (minutes)
First ISZ	10/4/57	TT	947	228	65.1	96.2
Second ISZ	11/3/57	TT	1,671	225	65.3	103.8
Third ISZ	5/15/58	TT	1,881	226	65.2	106.0
Luna 1	1/2/59	TT	1.31†	0.98†		450 D††
Luna 2	9/12/59	TT		(LUNAR IMPACT)		
Luna 3	10/4/59	TT	480,000	40,000	75.0	22,300.0
First space ship satellite	5/15/60	TT	369	312	65.0	91.2
Second space ship satellite	8/19/60	TT	339	306	65.0	90.7
Heavy satellite	12/1/60	TT	249	180	65.0	88.5

* According to data of The Bulletin of the Stations of Optical Observation of ISZ (Artificial Earth Satellites) of the Astronomical Council of the Academy of Sciences of the U.S.S.R.

**Launch site information provided by author.

† In Astronomical Units (AU), approximately 93 million miles.

††D: Days.

Notes: TT – Tyuratam/Baikonur
KY – Kapustin Yar
PL – Plesetsk

Satellite name	Date launched	Launch site	Apogee (kilometers)	Perigee (kilometers)	Inclination (degrees)	Period (minutes)
Venus satellite	2/4/61	TT	328	224	65.0	89.8
Heavy satellite	2/12/61	TT	1.02†	0.72†	65.0	300 D††
Heavy satellite	2/12/61	TT	282	229	65.0	89.6
Fourth space ship satellite	3/9/61	TT	249	184	65.0	88.6
Fifth space ship satellite	3/25/61	TT	247	178	49.0	88.4
Vostok	4/12/61	TT	327	181	65.0	89.3
Vostok 2	8/6/61	TT	244	183	65.0	88.5
Cosmos 1	3/16/62	KY	980	217	49.0	96.4
Cosmos 2	4/6/62	KY	1,546	212	49.0	102.3
Cosmos 3	4/24/62	KY	720	229	65.0	93.8
Cosmos 4	4/26/62	TT	330	298	65.0	90.6
Cosmos 5	5/28/62	KY	1,600	203	49.1	102.8
Cosmos 6	6/30/62	KY	360	274	49.0	90.6

Cosmos 7	7/28/62	TT	369	210	65.0	90.1
Vostok 3	8/11/62	TT	244	181	65.0	88.3
Vostok 4	3/12/62	TT	237	180	65.0	88.4
Cosmos 8	8/18/62	KY	609	256	49.0	92.9
Cosmos 9	9/27/62	TT	353	301	65.0	90.0
Cosmos 10	10/17/62	TT	380	210	65.0	90.2
Cosmos 11	10/20/62	KY	921	245	49.0	96.1
Mars 1	11/1/62	TT	(PASSED MARS AT 193,000 KM)			
Cosmos 12	12/22/62	TT	405	211	65.0	90.5
Cosmos 13	3/21/63	TT	337	205	65.0	89.8
Luna 4	4/2/63	TT	(PASSED MOON AT 8,500 KM)			
Cosmos 14	4/13/63	KY	512	265	49.0	92.1
Cosmos 15	4/22/63	TT	371	173	65.0	89.8
Cosmos 16	4/28/63	TT	401	207	65.0	90.4
Cosmos 17	5/22/63	KY	788	260	49.0	94.8
Cosmos 18	5/24/63	TT	301	209	65.0	89.4

Satellite name	Date launched	Launch site	Apogee (kilometers)	Perigee (kilometers)	Inclination (degrees)	Period (minutes)
Vostok 5	6/14/63	TT	222	175	65.0	88.3
Vostok 6	6/16/63	TT	231	181	65.0	88.3
Cosmos 19	8/6/63	KY	519	270	49.0	92.2
Cosmos 20	10/18/63	TT	311	206	65.0	89.6
Polyet 1	11/1/63	TT	1,437	339	58.9	102.5
Cosmos 21	11/11/63	TT	229	195	64.8	88.5
Cosmos 22	11/16/63	TT	394	205	64.9	90.3
Cosmos 23	12/13/63	KY	613	240	49.0	92.9
Cosmos 24	12/19/63	TT	408	211	65.0	91.0
Elektron 1	1/30/64	TT	7,100	406	61.0	169.0
Elektron 2	1/30/64	TT	68,200	460	61.0	1,360.0
Cosmos 25	2/27/64	KY	526	272	49.0	92.3
Cosmos 26	3/18/64	KY	403	271	49.0	91.0
Cosmos 27	3/27/64	TT	237	192	64.8	88.7

228

			(PASSED VENUS AT 100,000 KM)			
Zond 1	4/2/64	TT				
Heavy satellite	4/2/64	TT	213	187	64.8	88.5
Cosmos 28	4/4/64	TT	395	209	65.0	90.4
Polyet 2	4/12/64	TT	500	310	58.6	92.4
Cosmos 29	4/25/64	TT	309	204	65.1	89.5
Cosmos 30	5/18/64	TT	383	207	64.9	90.2
Cosmos 31	6/6/64	KY	508	228	49.0	91.6
Cosmos 32	6/10/64	TT	333	209	51.3	89.8
Cosmos 33	6/23/64	TT	293	209	65.0	89.4
Cosmos 34	7/1/64	TT	360	205	65.0	90.0
Elektron 3	7/11/64	TT	7,040	405	60.9	168.0
Elektron 4	7/11/64	TT	66,235	459	60.9	1,314.0
Cosmos 35	7/15/64	TT	268	217	51.3	89.2
Cosmos 36	7/30/64	TT	503	259	49.0	91.9
Cosmos 37	8/14/64	TT	300	205	65.0	89.5
Cosmos 38	8/18/64	TT	876	210	56.2	95.2

Satellite name	Date launched	Launch site	Apogee (kilometers)	Perigee (kilometers)	Inclination (degrees)	Period (minutes)
Cosmos 39	8/18/64	TT	876	210	56.2	95.2
Cosmos 40	8/18/64	TT	876	210	56.2	95.2
Cosmos 41	8/22/64	TT	34,855	394	64.0	715.0
Cosmos 42	8/22/64	KY	1,099	232	49.0	97.8
Cosmos 43	8/22/64	KY	1,099	232	49.0	97.8
Cosmos 44	8/28/64	TT	860	618	65.0	99.5
Cosmos 45	9/13/64	TT	327	206	64.9	89.7
Cosmos 46	9/24/64	TT	271	215	51.3	89.2
Cosmos 47	10/6/64	TT	413	177	64.8	90.0
Voskhod	10/12/64	TT	408	178	64.8	90.1
Cosmos 48	10/14/64	TT	295	203	65.1	89.4
Cosmos 49	10/24/64	KY	490	260	49.0	91.8
Cosmos 50	10/28/64	TT	241	196	51.3	88.7
Zond 2	11/30/64	TT	(PASSED MARS AT 1500 KM)			

Heavy satellite	11/30/64	TT	219	153	64.7	88.2
Cosmos 51	12/10/64	KY	554	264	48.8	92.5
Cosmos 52	1/11/65	TT	304	205	65.0	89.5
Cosmos 53	1/30/65	KY	1,192	227	48.8	98.7
Cosmos 54	2/21/65	TT	1,856	280	56.1	106.2
Cosmos 55	2/21/65	TT	1,856	280	56.1	106.2
Cosmos 56	2/21/65	TT	1,856	280	56.1	106.2
Cosmos 57	2/22/65	TT	695	175	64.8	91.1
Cosmos 58	2/26/65	TT	339	581	65.0	96.8
Cosmos 59	3/7/65	TT	339	209	65.0	89.7
Cosmos 60	3/12/65	TT	287	201	64.7	89.1
Cosmos 61	3/15/65	TT	1,837	273	56.0	106.0
Cosmos 62	3/15/65	TT	1,837	273	56.0	106.0
Cosmos 63	3/15/65	TT	1,837	273	56.0	106.0
Voskhod	3/18/65	TT	498	174	64.8	90.9
Cosmos 64	3/25/65	TT	271	206	65.0	89.2

Satellite name	Date launched	Launch site	Apogee (kilometers)	Perigee (kilometers)	Inclination (degrees)	Period (minutes)
Cosmos 65	4/17/65	TT	342	210	65.0	89.8
Molniya 1-A	4/23/65	TT	39,957	548	65.0	720.0
Cosmos 66	5/7/65	TT	291	197	65.0	89.3
Luna 5	5/9/65	TT	(LUNAR IMPACT)			
Heavy satellite	5/9/65	TT	217	151	64.8	88.3
Cosmos 67	5/25/65	TT	350	207	51.8	89.9
Luna 6	6/8/65	TT	(PASSED MOON AT 160,000 KM)			
Heavy satellite	6/8/65	TT	246	167	64.8	88.7
Cosmos 68	6/15/65	TT	334	205	65.0	89.8
Cosmos 69	6/25/65	TT	332	211	64.9	89.7
Cosmos 70	7/2/65	KY	1,154	229	56.0	98.4
Cosmos 71	7/16/65	TT	550	550	56.0	95.5
Cosmos 72	7/16/65	TT	550	550	56.0	95.5
Cosmos 73	7/16/65	TT	550	550	56.0	95.5

Cosmos 74	7/16/65	TT	550	550	56.0	95.5
Cosmos 75	7/16/65	TT	550	550	65.0	95.5
Proton 1	7/16/65	TT	627	190	56.0	92.5
Zond 3	7/18/65	TT	(PASSED MOON AT 9,220 KM)			
Heavy satellite	7/18/65	TT	210	164	56.0	88.2
Cosmos 76	7/23/65	KY	530	261	56.0	92.2
Cosmos 77	3/3/65	TT	291	200	56.0	89.3
Cosmos 78	8/14/65	TT	329	206	65.0	89.8
Cosmos 79	8/25/65	TT	359	211	64.9	90.0
Cosmos 80	9/3/65	TT	1,500	1,500	56.0	116.6
Cosmos 81	9/3/65	TT	1,500	1,500	56.0	116.6
Cosmos 82	9/3/65	TT	1,500	1,500	56.0	116.6
Cosmos 83	9/3/65	TT	1,500	1,500	56.0	116.6
Cosmos 84	9/3/65	TT	1,500	1,500	56.0	116.6
Cosmos 85	9/9/65	TT	319	212	65.0	89.6
Cosmos 86	9/18/65	TT	1,690	1,380	56.0	116.7

Satellite name	Date launched	Launch site	Apogee (kilometers)	Perigee (kilometers)	Inclination (degrees)	Period (minutes)
Cosmos 87	9/18/65	TT	1,690	1,380	56.0	116.7
Cosmos 88	9/18/65	TT	1,690	1,380	56.0	116.7
Cosmos 89	9/18/65	TT	1,690	1,380	56.0	116.7
Cosmos 90	9/18/65	TT	1,690	1,380	56.0	116.7
Cosmos 91	9/23/65	TT	342	212	65.0	89.8
Luna 7	10/4/65	TT	(LUNAR IMPACT)			
Heavy satellite	10/4/65	TT	286	129	64.8	88.6
Molniya 1-B	10/14/65	TT	40,000	500	65.0	71.9
Cosmos 92	10/16/65	TT	353	212	65.0	83.9
Cosmos 93	10/19/65	KY	522	220	48.4	91.7
Cosmos 94	10/28/65	TT	293	211	65.0	89.3
Proton 2	11/2/65	TT	637	191	63.5	92.6
Cosmos 95	11/4/65	KY	521	207	48.4	91.7
Venus 2	11/12/55	TT	1.08†	0.82†	4.3	366 D††

Object	Date						Category
Venus 3	11/16/65	TT	310	227	49.0	89.6	(VENUS IMPACT)
Cosmos 96	11/23/65	TT	2,100	220	65.0	108.3	
Cosmos 97	11/26/65	KY	570	216	65.0	92.0	
Cosmos 98	11/27/65	TT					
Luna 8	12/3/65	TT					(LUNAR IMPACT)
Heavy satellite	12/3/65	TT	209	169	65.0	88.2	
Cosmos 99	12/10/65	TT	320	199	65.0	89.6	
Cosmos 100	12/17/65	TT	650	650	49.0	97.7	
Cosmos 101	12/12/65	KY	550	260	65.0	92.4	
Cosmos 102	12/28/65	TT	278	218	56.0	89.2	
Cosmos 103	12/28/65	TT	600	600	65.0	97.0	
Cosmos 104	1/7/66	TT	401	204	65.0	90.2	
Cosmos 105	1/22/66	TT	324	204	48.4	89.7	
Cosmos 106	1/25/66	KY	564	290	48.4	92.8	
Luna 9	1/31/66	TT					(SOFT LUNAR LANDING)
Heavy satellite	1/31/65	TT	220	170	52.0	88.0	

Satellite name	Date launched	Launch site	Apogee (kilometers)	Perigee (kilometers)	Inclination (degrees)	Period (minutes)
Cosmos 107	2/10/66	TT	322	204	65.0	89.7
Cosmos 108	2/11/66	KY	865	227	48.9	95.3
Cosmos 109	2/19/66	TT	309	209	65.0	89.5
Cosmos 110	2/22/66	TT	904	187	51.9	95.3
Cosmos 111	3/1/66	TT	226	191	51.9	88.6
Cosmos 112	3/17/66	PL	565	214	72.0	92.1
Cosmos 113	3/21/66	TT	327	210	65.0	89.6
Luna 10	3/31/66	TT	(LUNAR ORBIT)			
Heavy satellite	3/31/66	TT	212	195	51.8	88.5
Cosmos 114	4/6/66	PL	374	210	73.0	90.1
Cosmos 115	4/20/66	TT	294	190	65.0	89.3
Molniya' 1-C	4/25/66	TT	39,500	499	64.5	710.0
Cosmos 116	4/26/66	KY	478	294	48.4	92.0
Cosmos 117	5/6/66	TT	308	207	65.0	89.5

Name	Date	Site				
Cosmos 118	5/11/66	TT	640	640	65.0	97.1
Cosmos 119	5/24/66	KY	1,305	219	48.5	99.8
Cosmos 120	6/8/66	TT	300	200	51.8	89.4
Cosmos 121	6/17/66	PL	354	210	72.9	89.9
Cosmos 122	6/25/66	TT	625	625	65.0	97.1
Proton 3	7/6/66	TT	630	190	63.5	92.5
Cosmos 123	7/8/66	KY	529	263	48.8	92.2
Cosmos 124	7/14/66	TT	303	208	51.8	89.4
Cosmos 125	7/20/66	TT	250	250	65.0	89.5
Cosmos 126	7/28/66	TT	359	212	51.8	90.0
Cosmos 127	8/8/66	TT	279	204	51.9	89.2
Luna 11	8/24/66	TT	(LUNAR ORBIT)			
Heavy satellite	8/24/66	TT	190	177	51.9	88.1
Cosmos 128	8/27/66	TT	364	212	65.0	90.0
Cosmos 129	10/14/65	PL	307	202	65.0	89.4
Molniya 1-D	10/20/66	TT	39,700	485	64.9	713.0

Satellite name	Date launched	Launch site	Apogee (kilometers)	Perigee (kilometers)	Inclination (degrees)	Period (minutes)
Cosmos 130	10/20/66	TT	340	211	65.0	89.8
Luna 12	10/20/66	TT		(LUNAR ORBIT)		
Heavy satellite	10/20/66	TT	212	199	51.9	88.6
Cosmos 131	11/12/66	PL	360	205	72.9	89.9
Cosmos 132	11/19/66	TT	280	207	65.0	89.3
Cosmos 133	11/28/66	TT	232	181	51.9	88.4
Cosmos 134	12/3/66	TT	319	214	65.0	89.6
Cosmos 135	12/12/66	KY	662	259	48.5	93.5
Cosmos 136	12/19/66	PL	305	198	64.6	89.4
Luna 13	12/21/66	TT		(LUNAR LANDING)		
Heavy satellite	12/21/66	TT	223	171	51.8	88.4
Cosmos 137	12/21/66	KY	1,720	230	48.8	104.3
Cosmos 138	1/19/67	PL	293	193	65.0	89.2
Cosmos 139	1/25/67	TT	210	144	50.0	88.0

Cosmos 140	2/7/67	TT	241	170	51.7	88.5
Cosmos 141	2/8/67	PL	345	210	72.9	89.8
Cosmos 142	2/14/67	KY	1,362	214	48.4	100.3
Cosmos 143	2/27/67	TT	302	204	65.0	89.5
Cosmos 144	2/28/67	PL	625	625	81.2	96.9
Cosmos 145	3/3/67	KY	2,135	220	48.4	108.6
Cosmos 146	3/10/67	TT	310	190	51.5	89.2
Cosmos 147	3/13/67	PL	317	198	65.0	89.5
Cosmos 148	3/16/67	PL	436	275	71.0	91.3
Cosmos 149	3/21/67	KY	297	248	48.4	89.8
Cosmos 150	3/22/67	PL	373	206	65.7	90.1
Cosmos 151	3/24/67	TT	630	630	56.0	97.1
Cosmos 152	3/25/67	PL	512	283	71.0	92.2
Cosmos 153	4/4/67	PL	291	202	64.6	89.3
Cosmos 154	4/8/67	TT	232	186	51.6	88.5
Cosmos 155	4/12/67	TT	286	203	51.8	89.2

Satellite name	Date launched	Launch site	Apogee (kilometers)	Perigee (kilometers)	Inclination (degrees)	Period (minutes)
Soyuz 1	4/23/67	TT	224	201	51.7	88.6
Cosmos 156	4/27/67	PL	630	630	81.2	97.0
Cosmos 157	5/12/67	TT	296	202	51.3	89.4
Cosmos 158	5/15/67	PL	850	850	74.4	100.7
Cosmos 159	5/16/67	TT	60,600	380	51.8	1,173.0
Cosmos 160	5/17/67	TT	205	142	49.6	88.0
Cosmos 161	5/22/67	PL	343	205	65.7	89.8
Molniya 1-E	5/24/67	TT	39,810	460	64.8	715.0
Heavy satellite	5/25/67	TT	312	203	67.9	91.3
Cosmos 162	6/1/67	TT	280	201	51.8	89.2
Cosmos 163	6/5/67	KY	616	261	48.4	93.1
Cosmos 164	6/8/67	PL	320	202	65.7	89.5
Venus 4	6/12/67	TT	(VENUS LANDING)			
Heavy satellite	6/12/67	TT	188	162	51.8	87.9

Cosmos 165	5/12/67	PL	1,542	211	81.9	102.1
Cosmos 166	6/16/67	KY	578	283	48.4	92.9
Cosmos 167	6/17/67	TT	286	201	51.8	89.2
Cosmos 168	7/4/67	TT	268	199	51.8	89.1
Cosmos 169	7/17/67	TT	208	144	50.0	88.8
Cosmos 170	7/31/67	TT	208	145	50.0	88.0
Cosmos 171	8/8/67	TT	220	145	50.0	88.0
Cosmos 172	8/9/67	TT	301	202	51.8	89.4
Cosmos 173	8/24/67	PL	528	280	71.0	92.3
Cosmos 174	8/31/67	TT	29,750	500	64.5	715.0
Cosmos 175	9/11/67	PL	386	210	72.9	92.2
Cosmos 176	9/12/67	PL	1,581	206	81.9	102.5
Cosmos 177	9/16/67	TT	292	202	51.8	89.3
Cosmos 178	9/19/57	TT	205	145	50.0	88.0
Cosmos 179	9/22/67	TT	208	145	50.0	88.0
Cosmos 180	9/26/67	PL	370	212	72.9	90.1
Molniya 1-F	10/3/67	TT	39,600	465	65.0	712.0

Satellite name	Date launched	Launch site	Apogee (kilometers)	Perigee (kilometers)	Inclination (degrees)	Period (minutes)
Cosmos 181	10/11/67	PL	706	664	99.2	101.1
Cosmos 182	10/16/67	TT	353	209	65.0	89.9
Cosmos 183	10/18/67	KY	212	145	49.0	89.9
Molniya 7-G	10/22/67	TT	39,600	465	64.7	71.2
Cosmos 184	10/24/67	PL	635	635	81.2	97.1
Cosmos 185	10/27/67	TT	888	522	64.1	98.7
Cosmos 186	10/27/67	TT	234	208	51.7	88.7
Cosmos 187	10/28/67	TT	210	145	50.0	87.8
Cosmos 188	10/30/67	TT	276	200	51.7	89.0
Cosmos 189	10/30/67	PL	600	535	74.0	95.7
Cosmos 190	11/3/67	TT	347	201	65.7	89.8
Cosmos 191	11/21/67	PL	518	281	71.0	92.2
Cosmos 192	11/23/67	PL	755	755	74.0	99.9
Cosmos 193	11/25/67	PL	352	204	65.7	89.9

	Date	Site				
Cosmos 194	12/3/67	PL	332	204	65.7	89.7
Cosmos 195	12/16/67	PL	373	210	65.7	90.1
Cosmos 196	12/19/67	KY	884	224	45.0	95.5
Cosmos 197	12/26/67	KY	502	219	48.5	91.5
Cosmos 198	12/27/67	TT	174 to 1,000	164 to 556	65.1	89.8
Cosmos 199	1/16/68	PL	386	204	65.7	90.2
Cosmos 200	1/19/68	PL	536	536	74.0	95.2
Cosmos 201	2/6/68	TT	355	210	65.0	89.9
Cosmos 202	2/20/68	KY	502	220	48.4	91.5
Cosmos 203	2/20/63	PL	1,200	1,200	74.1	109.4
Zond 4	3/2/68	TT	206	190	51.5	88.4
Cosmos 204	3/5/68	PL	873	282	71.0	95.9
Cosmos 205	3/5/68	PL	310	201	65.7	89.4
Cosmos 206	3/14/68	PL	630	630	81.0	97.0
Cosmos 207	3/16/68	PL	342	210	65.6	89.8
Cosmos 208	3/21/68	TT	305	207	65.0	89.4

Satellite name	Date launched	Launch site	Apogee (kilometers)	Perigee (kilometers)	Inclination (degrees)	Period (minutes)
Cosmos 209	3/22/68	TT	282	250	65.1	89.6
Cosmos 210	4/3/68	PL	395	217	81.2	90.3
Luna 14	4/7/68		(LUNAR ORBIT)			
Cosmos 211	4/9/68	PL	1,574	210	81.9	102.5
Cosmos 212	4/14/68	TT	239	210	51.7	88.8
Cosmos 213	4/15/68	TT	291	205	51.4	89.2
Cosmos 214	4/18/68	PL	403	211	81.4	90.3
Cosmos 214	4/18/68	KY	426	261	48.5	91.1
Cosmos 216	4/20/68	TT	277	199	51.8	89.1
Molniya 1-H	4/21/68	TT	39,700	460	65.0	713.0
Cosmos 217	4/24/68	TT	520	396	62.2	93.4
Cosmos 218	4/25/68	TT	210	144	50.0	87.8
Cosmos 219	4/26/68	KY	1,770	222	48.4	104.7
Cosmos 220	5/7/68	PL	760	670	74.0	99.2

Cosmos 221	5/24/68	KY	2,108	220	48.4	108.3
Cosmos 222	5/30/68	PL	528	277	71.0	92.3
Cosmos 223	6/1/68	PL	374	212	72.9	90.1
Cosmos 224	6/4/68	TT	270	200	51.8	89.0
Cosmos 225	6/11/68	KY	530	257	48.4	92.2
Cosmos 226	6/12/68	PL	650	603	81.2	96.9
Cosmos 227	6/18/68	TT	281	194	51.8	89.1
Cosmos 228	6/21/68	TT	259	206	51.6	89.0
Cosmos 229	6/26/68	PL	354	210	72.8	88.9
Cosmos 230	7/5/68	KY	580	290	48.5	93.0
Molniya 1-J	7/6/68	TT	39,770	470	65.0	715.0
Cosmos 231	7/10/68	TT	330	211	65.0	89.7
Cosmos 232	7/16/68	PL	352	202	65.0	89.8
Cosmos 233	7/18/68	PL	1,545	210	82.0	102.1
Cosmos 234	7/30/68	TT	310	210	51.8	89.5
Cosmos 235	8/9/68	TT	303	207	51.8	89.4

Satellite name	Date launched	Launch site	Apogee (kilometers)	Perigee (kilometers)	Inclination (degrees)	Period (minutes)
Cosmos 236	8/27/68	TT	655	600	56.0	96.9
Cosmos 237	8/27/68	PL	343	201	64.5	89.7
Cosmos 238	8/28/68	TT	219	199	51.7	88.5
Cosmos 239	9/5/68	TT	282	202	51.8	89.2
Cosmos 240	9/14/68	TT	293	197	51.8	89.3
Zond 5	9/14/68	TT	(CIRCUMLUNAR FLIGHT)			
Cosmos 241	9/16/68	PL	343	201	65.4	89.7
Cosmos 242	9/20/68	PL	440	280	71.0	91.3
Cosmos 243	9/23/68	TT	319	210	71.3	89.6
Cosmos 244	10/2/68	KY	212	140	50.0	
Cosmos 245	10/3/68	PL	509	282	71.0	92.1
Molniya 1-K	10/5/68	TT	39,600	490	65.0	712.0
Cosmos 246	10/7/68	PL	348	147	65.4	89.4
Cosmos 247	10/11/68	PL	362	205	65.4	89.9

Name	Date	Type				
Cosmos 248	10/19/68	TT	551	490	62.3	94.8
Cosmos 249	10/20/68	TT	2,177	514	62.4	112.2
Soyuz 2	10/25/68	TT	224	185	51.7	88.5
Soyuz 3	10/26/68	TT	225	205	51.4	88.6
Cosmos 250	10/30/68	PL	556	523	74.0	95.3
Cosmos 251	10/31/68	TT	270	198	65.0	89.1
Cosmos 252	11/1/68	TT	2,172	538	61.9	112.5
Zond 6	11/10/68	TT	(CIRCUMLUNAR FLIGHT)			
Cosmos 253	11/13/68	PL	335	206	65.4	89.9
Proton 4	11/16/68	TT	495	255	51.5	91.8
Cosmos 254	11/21/68	PL	350	203	65.4	89.8
Cosmos 255	11/29/68	PL	336	201	65.4	89.7
Cosmos 256	11/30/68	PL	1,234	1,168	74.1	109.3
Cosmos 257	12/3/68	PL	470	282	71.0	91.1
Cosmos 258	12/10/68	TT	325	210	65.0	89.6
Cosmos 259	12/14/68	KY	1,353	219	48.5	100.3

Satellite name	Date launched	Launch site	Apogee (kilometers)	Perigee (kilometers)	Inclination (degrees)	Period (minutes)
Cosmos 260	12/16/68	TT	39,600	500	65.0	712.0
Cosmos 261	12/19/68	PL	670	217	71.0	93.1
Cosmos 262	12/26/68	KY	818	263	48.5	95.2
Venus 5	1/5/69	TT		(VENUS LANDING)		
Venus 6	1/10/69	TT		(VENUS LANDING)		
Cosmos 263	1/12/69	PL	346	205	65.4	89.8
Soyuz 4	1/14/69	TT	225	173	51.7	88.3
Soyuz 5	1/15/69	TT	230	200	51.7	88.7
Cosmos 264	1/23/69	TT	330	219	70.0	89.7
Cosmos 265	2/7/69	PL	485	283	71.0	91.9
Cosmos 266	2/25/69	PL	358	208	72.9	89.9
Cosmos 267	2/26/69	TT	346	210	65.0	89.9
Cosmos 268	3/5/69	KY	2,186	219	48.4	109.2
Cosmos 269	3/5/69	PL	558	526	74.0	95.3

Cosmos 270	3/6/69	PL	350	205	65.4	89.8
Cosmos 271	3/15/69	PL	342	200	65.4	89.7
Cosmos 272	3/17/69	PL	1,220	1,195	74.0	109.4
Cosmos 273	3/22/69	PL	356	205	65.4	89.9
Cosmos 274	3/24/69	TT	323	213	65.0	89.6
Meteora 1	3/26/69	PL	713	644	81.2	97.9
Cosmos 275	3/28/69	PL	805	248	71.0	95.2
Cosmos 276	4/4/69	PL	410	214	81.4	90.4
Cosmos 277	4/4/69	PL	494	280	71.0	92.0
Cosmos 278	4/9/69	PL	338	203	65.0	89.7
Molniya 1-L	4/11/69	TT	39,700	470	65.0	713.0
Cosmos 279	4/15/69	TT	280	194	51.8	89.1
Cosmos 280	4/23/69	TT	272	206	51.6	89.1
Cosmos 281	5/13/69	PL	317	194	65.4	89.3
Cosmos 282	5/20/69	PL	343	209	65.4	89.8
Cosmos 283	5/27/69	PL	1,539	210	82.0	102.1

Satellite name	Date launched	Launch site	Apogee (kilometers)	Perigee (kilometers)	Inclination (degrees)	Period (minutes)
Cosmos 284	5/29/69	TT	308	207	51.8	89.5
Cosmos 285	6/3/69	PL	518	279	71.0	92.2
Cosmos 286	6/15/69	PL	349	206	65.4	89.8
Cosmos 287	6/24/69	TT	268	190	51.8	89.0
Cosmos 288	6/27/69	TT	281	201	51.8	89.2
Cosmos 289	7/10/69	PL	350	200	65.4	89.8
Luna 15	7/13/69	TT	(LUNAR ORBIT AND LANDING)			
Cosmos 290	7/22/69	PL	352	200	65.4	89.9
Molynia 1-M	7/22/69	TT	39,540	520	64.9	711.0
Cosmos 291	8/6/69	TT	574	153	62.3	91.5
Zond 7	8/7/69	TT				
Cosmos 292	8/13/69	PL	786	747	74.0	99.9
Cosmos 293	8/16/69	TT	270	208	51.8	89.1
Cosmos 294	8/19/69	PL	348	202	65.4	89.8

Cosmos 295	8/22/69	PL	500	282	71.0	92.0
Cosmos 296	8/29/69	TT	322	211	65.0	89.6
Cosmos 297	9/2/69	PL	334	211	72.9	89.7
Cosmos 298	9/15/69	TT	212	140	50.0	
Cosmos 299	9/18/69	TT	311	214	65.0	89.5
Cosmos 300	9/23/69	TT	208	190	51.5	88.2
Cosmos 301	9/24/69	PL	307	197	65.4	89.4
Meteora 2	10/6/69	PL	690	630	81.2	97.7
Soyuz 6	10/11/69	TT	223	186	51.7	88.4
Soyuz 7	10/12/69	TT	226	207	51.7	88.6
Soyuz 8	10/13/69	TT	223	205	51.7	88.6
Intercosmos 1	10/14/69	KY	640	260	48.4	93.3
Cosmos 302	10/17/69	PL	340	202	65.4	89.7
Cosmos 303	10/18/69	PL	492	282	71.0	91.9
Cosmos 304	10/21/69	PL	774	747	74.0	99.9
Cosmos 305	10/22/69	TT	205	193	51.5	

Satellite name	Date launched	Launch site	Apogee (kilometers)	Perigee (kilometers)	Inclination (degrees)	Period (minutes)
Cosmos 306	10/24/69	TT	332	208	65.0	89.7
Cosmos 307	10/24/69	KY	2,178	220	48.4	109.1
Cosmos 308	11/4/69	PL	422	281	71.0	91.3
Cosmos 309	11/12/69	PL	384	203	65.4	90.1
Cosmos 310	11/15/69	TT	347	208	65.0	89.8
Cosmos 311	11/24/69	PL	496	284	71.0	92.0
Cosmos 312	11/24/69	PL	1,187	1,145	74.0	108.6
Cosmos 313	12/3/69	PL	276	204	65.4	89.1
Cosmos 314	12/11/69	PL	491	282	71.0	91.9
Cosmos 315	12/20/69	PL	556	521	74.0	95.3
Cosmos 316	12/23/69	KY	1,650	154	49.5	102.7
Cosmos 317	12/23/69	PL	302	209	65.4	89.4
Intercosmos 2	12/26/69	KY	640	260	48.4	93.3

APPENDIX B
A Comparison–
The First Ten Years
(1957–1967)

Table B-1
Chronology of Yearly Space Launches

Year	UNITED STATES Successful launches	Failures	Successes	SOVIET UNION Successful launches	Failures	Successes
1957	0	0	0	2	0	2
1958	7	2	5	1	0	1
1959	11	6	5	3	1	2
1960	16	3	13	5	4	1
1961	29	8	21	6	2	4
1962	52	5	47	20	6	14
1963	37	0	37	17	2	15
1964	54	0	54	30	1	29
1965	62	1	61	48	8	40
1966	70	5	65	44	4	40
1967	50	2	48	63	3	60
Totals	**388**	**32**	**356**	**239**	**31**	**208**

Table B-2
Chronology of Unmanned Space Firsts

No.	Year	Date	NATIONAL EVENT United States	NATIONAL EVENT Soviet Union
1	1957	Oct. 4		Earth satellite (Sputnik 1)
2		Nov. 3		Dog Laika in orbit (Sputnik 2)
3	1958	Jan. 31	Van Allen belts (Explorer 1)	
4		Mar. 17	Pear-shaped earth (Vanguard 1)	
5		Oct. 11	Deep-space probe (Pioneer 1)	
6		Dec. 18	Passive Comsat (Score)	
7	1959	Feb. 28	Polar orbit (Discoverer 1)	
8		July 7	Solar panels	
9		Sept. 14		Lunar impact (Lunik 2)
10		Oct. 7		Lunar far-side photography (Lunik 3)
11	1960	Mar. 11	Radio telescope commands (Pioneer 5)	
12		Apr. 1	Metsat (Tiros 1)	
13		Apr. 13	Navsat (Transit 1B)	
14		May 24	ICBM alarm (Midas 2)	
15		Aug. 10	Orbital recovery (Discoverer 13)	
16		Aug. 12	Passive Comsat (Echo 1)	
17		Aug. 20		Dogs returned live from orbit (Sputnik 5)
18		Oct. 4	Active repeater Comsat (Courier 1B)	
19	1961	June 29	Triple payload launch (Transit 4A)	

Table B-2 (Continued)

No.	Year	Date	NATIONAL EVENT United States	NATIONAL EVENT Soviet Union
20			SNAP power source (Transit 4A)	
21	1962	Mar. 7	Earth solar observatory	
22		July 10	Commercial communication satellite	
23		Oct. 31	Geodetic satellite	
24		Nov. 1		Mars probe
25	1963	July 26	Synchronous satellite	
26		Nov. 1		Polet 1—Maneuvering spacecraft
27		Nov. 26	Interplanetary magnetic fields	
28	1964	Dec. 8–18		Electrojet plasma engines-control
29	1965	Mar. 21	Live TV of lunar impact	
30	1965	July 14	Photos of Mars surface	
31	1966	Feb. 3		Soft lunar landing
32		Apr. 3		Lunar orbiter
33		Aug. 8	Photo of earth from moon	
34	1967	Feb. 5	Surveyor 1, photographed lunar orbiter	
35		May 4	Polar lunar orbiter	
36		June 6		Venus 4 lander
37		Oct. 30		Unmanned rendezvous/docking

Table B-3
Chronology of Manned Time in Orbit

No.	Year	Date	NATIONAL CUMULATIVE MANHOURS			
			United States		Soviet Union	
1	1961	Apr. 12			Gagarin	1:48
2		May 5	Shepard	0:15		
3		July 21	Grissom*	0:31		
4		Aug. 6			Titov	27:06
5	1962	Feb. 20	Glenn	5:26		
6		May 24	Carpenter	10:22		
7		Aug. 11			Nikolayev	121:28
8		Aug. 12			Popovich	192:25
9		Oct. 3	Schirra*	19:35		
10	1963	May 15	Cooper*	53:55		
11		June 14			Bykovsky	311:31
12		June 16			Tereshkova	382:21
13	1964	Oct. 12			Komarov*, Yegorov, Feoktistov	455:12
14	1965	Mar. 18			Belyayev, Leonov	507:16
15		Mar. 23	Grissom*, Young*	63:41		
16		June 3	McDivitt, White	259:33		
17		Aug. 21	Cooper*, Conrad*	641:25		
18		Dec. 4	Borman, Lovell*	1302:35		
19		Dec. 15	Schirra*, Stafford*	1354:17		
20	1966	Mar. 16	Armstrong, Scott	1375:39		
21		June 6	Stafford*, Cernan	1520:21		
22		July 18	Young*, Collins	1661:55		
23		Sept. 12	Conrad*, Gordon	1804:29		
24		Nov. 11	Lovell*, Aldrin	1993:37		
25	1967	Apr. 23			Komarov*	532:53

* Men who have gone into space twice.

258

Table B-4
Chronology of Manned Spaceflight Records

| No. | Year | Date | NATIONAL EVENT | |
			United States	Soviet Union
1	1958–1967		Successful launches, 388	
2	1961–1966		Successful manned launches, 16	
3	1961–1966		Total manned flights, 26	
4	1961–1966		Manhours in space, 1993.37	
5	1963	June 14		Duration—Single-man flight by Bykovsky, 119.0 hr.
6		June 16		Duration—Single-woman flight by Tereshkova, 70.50 hr.
7	1964	Oct. 12		Duration—Three-man flight by Komarov, Yegorov, Feoktistov, 24.17 hr.
8		Dec. 4	Duration—Two-man flight by Borman & Lovell, 330.35 hr.	
9	1965–1966		Total rendezvous flights, 10	
10	1965–1966		Total docking flights, 9	
11	1965–1966		Total EVA flights, 5	
12	1965–1966		Duration single EVA, 5 hr., 58 min.	
13	1965–1966		Duration—EVA total 12 hr., 39 min.	
14	1965–1966	Sept. 12	Highest altitude, Conrad & Gordon, 851 mi.	

Table B-5
Chronology of Manned Space Firsts

No.	Year	Date	NATIONAL EVENT	
			United States	Soviet Union
1	1961	April 12		Yuri Gagarin in orbit
2	1962	Aug. 12–15		Formation flight (Nikolayev & Popovich)
3	1963	Jun. 16		V. Tereshkova—woman cosmonaut
4	1964	Oct. 12		Multiplace spaceship (Komarov, Yegorov, & Feoktistov)
5	1965	Mar. 18		Emergence into space (Leonov)
6		Mar. 23	Two-man flight (Grissom & Young)	
7		Mar. 23	Piloted orbital plane changes	
8		June 3	EVA-controlled movements in space (White)	
9		Dec. 15	Piloted rendezvous (Shirra & Stafford)	
10	1966	Mar. 18	Spacecraft docking (Armstrong & Scott)	
11		July 18	Crewman touches another S/C (Collins)	
12		July 18	Post-docking maneuvers ⎤ Young	
13		July 18	Maneuvers with T/V prop. & prop. ⎦ & Collins	
14		Sept. 12	Station keeping; 100-ft nylon tether (Conrad & Gordon)	

Table B-6
Comparison of Biological Data Orbital Flights[a]

Year	United States	U.S.S.R.
1957	0	1
1958	0	1
1959	0	1
1960	0	3
1961	1	7
1962	4	9
1963	5	11
1964	5	12
1965	10	13
1966	15	14
1967	15	15

[a] Cumulative totals.

Table B-7
Comparison of Lunar Launches[a]

Year	United States		U.S.S.R.	
	Total Launches	Total Successes	Total Launches	Total Successes
1957	—	—	—	—
1958	—	—	—	—
1959	1	0	3	2
1960	1	0	3	2
1961	1	0	3	2
1962	4	0	3	2
1963	4	0	4	2
1964	4	1	4	2
1965	8	3	4	3
1966	13	6	14	8
1967	21	13	14	8

[a] Cumulative totals.

Table B-8
Comparison of Planetary Launches[a]

Year	United States Total Launches	United States Total Successes	U.S.S.R. Total Launches	U.S.S.R. Total Successes
1957	—	—	—	—
1958	—	—	—	—
1959	—	—	—	—
1960	—	—	—	—
1961	—	—	—	—
1962	Venus—1	Venus—0	Mars—2	Mars—0
1963	1	0	2	0
1964	Mars—3	2	2	0
1965	3	2	Venus—6	0
1966	3	2	6	0
1967	Venus—4	3	Venus—7	1

[a] Cumulative totals.

Bibliography

Anoschchenko, N. D., ed. *History of Aviation and Cosmonautics.* Vol. 1. Moscow: Academy of Sciences of the U.S.S.R., National Association of Historians of Natural Science and Technology, 1964.

"A.T.D. Press." Washington, D.C.: Translations by the Aerospace Technology Division, Library of Congress.

Blagonravov, A. A. *Soviet Rocketry: Some Contributions to Its History.* Moscow: Academy of Sciences of the U.S.S.R., National Association of Historians of Natural Science and Technology, 1964.

Branigan, T. L., ed. *Space Log.* Redondo Beach, Cal.: A quarterly publication of T.R.W. Space Technology Laboratories. Information drawn from data released by the National Aeronautics and Space Administration, Department of Defense, United Nations, Tass, and other official sources.

Cox, D., and M. Stoiko. *Spacepower, What It Means to You.* Philadelphia: The John C. Winston Company, Inc., 1958.

Krieger, F. J. *Casebook on Soviet Astronautics.* Parts 1 and 2. Santa Monica, Cal.: The Rand Corporation, 1956.

Kullas, A. J. *Duress for Technology.* New York: American Institute of Aeronautics and Astronautics, Paper No. 68-1106, 1968.

Parry, A. *Russia's Rockets and Missiles.* New York: Doubleday and Company, 1960.

Sheldon, C. S. *Review of the Soviet Space Program.* Washington, D.C.: U.S. Government Printing Office, 1967.

Slukhai, I. A. *Russian Rocketry, A Historical Survey.* Moscow: Academy of Science of the U.S.S.R., National Association of Historians of Natural Science and Technology, 1965.

Sokolsky, V. N. *A Short Outline of the Development of Rocket Research in the U.S.S.R.* Moscow: Academy of Sciences of the U.S.S.R., 1960. Translated by the U.S. Department of Commerce, Federal Scientific and Technical Information, Cat. No. 1924.

Sokolsky, V. N. *Russian Solid-Fuel Rockets.* Moscow: Academy of Sciences of the U.S.S.R., Institute of the History of Science and Engineering, 1963.

"Soviet-Bloc Research in Geophysics, Astronomy and Space." Washington, D.C.: Translations by the U.S. Department of Commerce, Federal Scientific and Technical Information.

Soviet Space Programs, 1962–1965; Goals and Purposes Achievements, Plans and International Implications. Washington, D.C.: U.S. Government Printing Office, 1966.

Soviet Space Programs: Organization, Plans, Goals and International Implications. Washington, D.C.: U.S. Government Printing Office, 1962.

Tokaty, G. A. "Foundations of Soviet Cosmonautics." *Spaceflight,* London: The British Interplanetary Society, 1967.

Tokaty, G. A. "Soviet Space Technology." *Spaceflight,* London: The British Interplanetary Society, 1963.

Tsander, F. A. "Problems of Flight by Jet Propulsion." *Collection of Articles, L. K. Koroev,* ed. Moscow: Publishing House of the Defense Industry, 1961.

Tsiolkovsky, K. E. *Works on Rocket Technology.* Moscow: Publishing House of the Defense Industry, 1947.

Wade, D. K. E. *Public Diplomacy and the Politics of the Space Race.* Washington, D.C.: The Fletcher School of Law and Diplomacy. 1967.

Index